THE ART CALCULUS

AN INTRODUCTION TO CALCULUS

Kenneth R. Williams

INSPIRATION BOOKS

Published by Inspiration Books, 2015
Kensglen, Nr Carsphairn, Castle Douglas, DG7 3TE, Scotland, U.K.

2nd edition 2015

http://www.vedicmaths.org

ISBN: 978-1-902517-40-7

CONTENTS

PREFACE

The aim of this book is to introduce Calculus in a simple holistic way that even young children can assimilate. It is a first introduction to Calculus and can be used in different ways.

* Prerequisites are: basic graph work, an understanding of what a gradient is and how to measure it (though this is briefly explained in the text) and a basic knowledge of algebra.

* Calculus is normally taught later in the school career, but here you will see that this does not have to be the case.

* Calculus can be taught and understood without the complex terminology and symbolism that is usually associated with this subject.

* The approach described here allows the possibility of connecting at any point with the conventional approach.

* So the simple Calculus methods shown are also connected in this book to the standard notations.

* The teacher/parent can therefore choose the appropriate material and introduce standard notations as appropriate for their class/child.

Possible paths through the book.

Without understanding formal algebra a good introduction to Calculus is still possible because children naturally generalise but may be puzzled by the use of letters to express those generalisations. So taking only the essential ideas from Chapters 1, 2, 3, 9, 10 a grounding in Calculus is entirely possible. (Chapter 5 can also be used by extending the idea of Chapter 3).

Path A Chapters: **1** **2** **3** - (5) - - - **9** **10** - - - -

Path B would be for a child who knows basic algebra as well as gradients but needs a gentle introduction. This child may go through Chapters 1, 2, 3, 4, 5, 9, 10.

Path B Chapters: **1** **2** **3** **4** **5** - - - **9** **10** - - (13) (14)

Path C would go through the entire book.

Path C Chapters: **1 2 3 4 5 6 7 8 9 10 11 12 13 14**

Chapters 6, 7, 8 deal with negative and fractional powers so may be felt a little harder (though the text does explain what is needed), but the beautiful graphical designs here are very attractive and easy to do.

Chapters 11, 12 introduce the integration symbol and are really present to connect the approach of this book to the traditional approach.

Chapters 4 and 13 are applications of Calculus, and Chapter 14 is an easy introduction to simple differential equations.

This book aims to open up the field of Calculus in a natural way, so that even young children can appreciate and apply the techniques in their thinking, in their study of mathematics and in the solution of problems.

INTRODUCTION

Calculus is all about change and growth. It is a way of measuring change and growth. And since these are everywhere it is no real surprise that Calculus has applications in very many areas: economic, medical, engineering and so on.

All growth is lawful, each has its own way of changing. And these laws of growth can be treated mathematically, with Calculus, so that we can make predictions about future behaviour or find out the laws of growth from growth patterns.

Nature is complex on the surface, and changes lead to more change: one thing affects another, and in particular one thing sets limits on another. So change inevitably leads to more change and limits of change.

The tree is blown by the wind, and if its root system is not strong enough, it may get blown over. A limit was reached. These limits are seen everywhere and are at the basis of Calculus.

DETAILED CONTENTS

CHAPTER 1

Growth and Limits

In this Chapter:

*** We look at different types of growth.**
*** How limits can be used in mathematics.**
*** What is meant by the gradient of a curve**

Calculus is all about growth and change. It is a mathematical way to describe change and therefore make predictions about how things will be in the future, or about how they were in the past.

Since growth and change are everywhere calculus has a vast range of uses as you will see.

GROWTH

We see growth all around us. And there are many types of growth.

When things grow they get bigger. And we may think of things that get smaller as having negative growth.

ZERO GROWTH

Suppose a young tree, a sapling, has a height of 10 cm.

During the winter it may not grow at all. If a graph was drawn of its height against time it would show a horizontal line, as shown next.

This graph shows that the height remains constant at 10 cm. We say there is no growth and the gradient (or slope) of the line is zero.

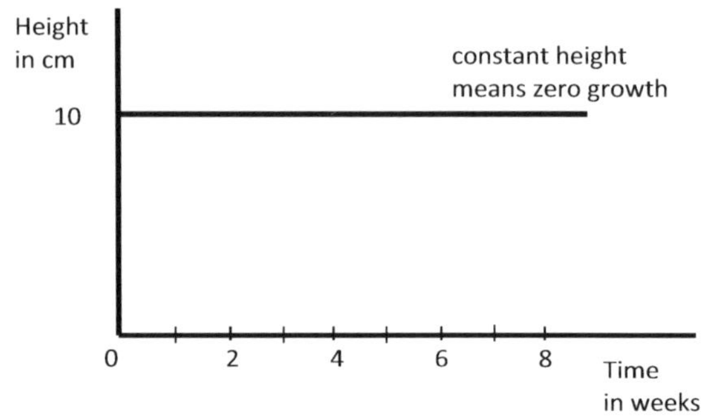

STEADY GROWTH

When the tree does start to grow the growth may be constant, meaning that each week its height increases by a constant amount.

Suppose it grows at 5 cm per week.

The graph would look like this:

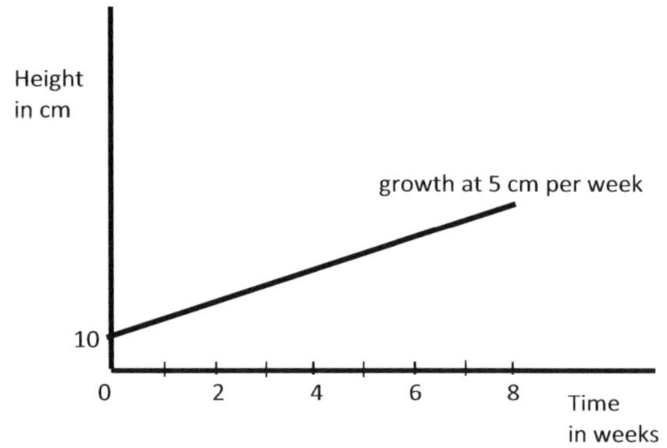

Here the line slopes upwards, indicating growth.
And because the growth is constant the line is straight.

How would the graph be different if the tree was growing at a faster rate? Suppose it is growing at 10 cm per week instead of 5.

Well, the line would still be straight, but it would be steeper, like this:

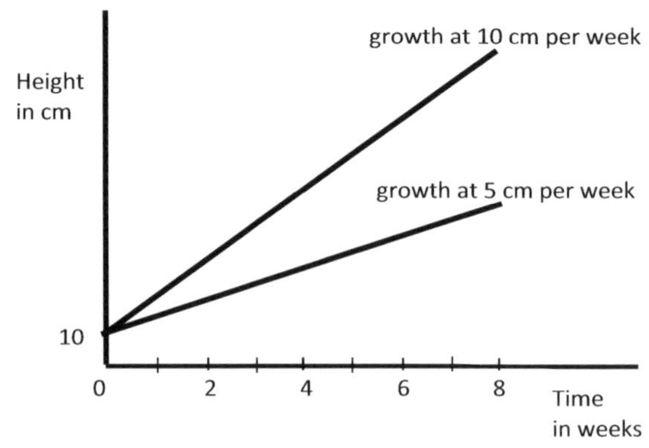

We can see that faster growth means steeper lines on the graph.
So the gradient of the upper line is greater than the gradient of the lower line.

Gradient, by definition, means the vertical increase of the line corresponding to an increase of 1 unit in the horizontal direction.

This means the gradient is the rate of growth: 5 cm per week, 10 cm per week, etc.

Exercise A (Answers Page 126)

a A sapling is 15 cm high and grows at 3 cm per week for 4 weeks. What is its height after 4 weeks?

b Draw a sketch showing this information.

VARIABLE GROWTH

But in practice the growth would not be constant. Through spring and summer the days are getting warmer and the tree grows faster as time goes on.

How would this look on the graph?

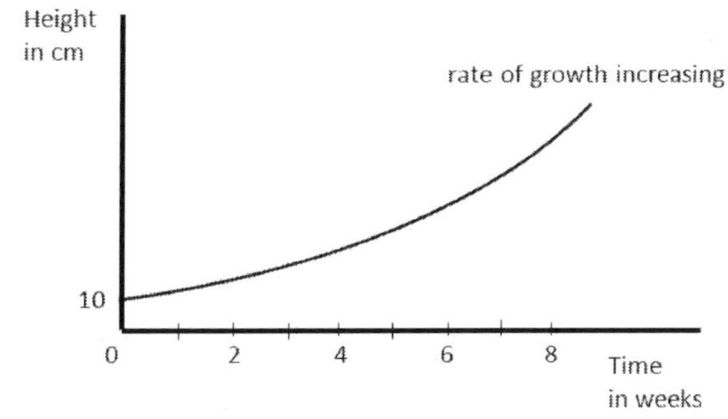

Well, we would see that the gradient is gradually increasing, as shown opposite:

The line is getting steeper.

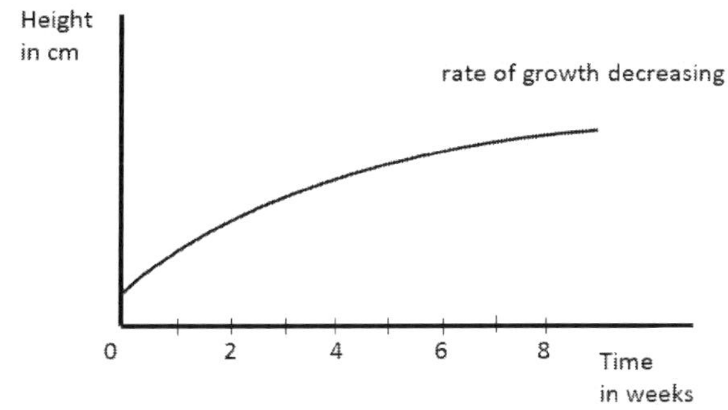

Later in the year the rate of growth will be diminishing and that may look like this:

Note how the gradient is getting less as time goes on.

LIMITS

Growth and limits are closely connected. If you throw a ball upwards its height will be growing, and there will be a limit to how high it can go.

Limits are everywhere: there are age limits, speed limits, time limits. There is a limit to how fast you can run, how much you can lift, how tall you will grow.

In fact everything has its boundaries, which set its limits. An organisation is limited by the rules that govern it, by its finances and the people who operate within it.

In mathematics, limits come up in all kinds of ways, and they are an extremely useful and powerful tool.

A limit may or may not be reached, as you will see.

EXAMPLE 1 - a spider

A small spider is 2 m from a wall and is crawling towards the wall. It takes the spider one hour to cover the first metre. But then it's getting tired and covers only half a metre in the next hour. Still getting tired it covers only a quarter of a metre in the next hour, and in fact always covers half the previous distance in every hour.

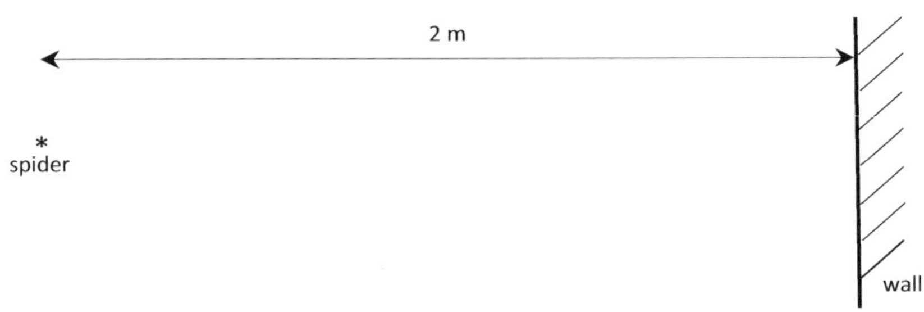

Will the spider ever reach the wall?

The spider will never reach the wall because at the end of each hour there will always be some distance left to cover.

We say the wall is the limit, which the spider will never reach, but which it may get closer and closer to.

EXAMPLE 2 - fractions

Suppose we need to decide which is greater: $\frac{4}{7}$ or $\frac{5}{8}$.

Noticing that the numerators, 4 and 5, are consecutive and the denominators, 7 and 8, are consecutive, we can continue the sequence to the right like this:

$$\frac{4}{7} \quad \frac{5}{8} \quad \frac{6}{9} \quad \frac{7}{10} \quad \cdots$$

Write out a few more fractions in the sequence.
If the numerator was 100, what would the denominator be?
You may notice that the denominator of each fraction is always 3 more than the numerator, so the fraction would be: $\frac{100}{103}$.

Is the sequence heading towards a limit?

Well we know that $\frac{100}{100} = 1$, so $\frac{100}{103}$ must be close to 1.

And as we go further along in the sequence we get closer and closer to 1.

So the limit of the original sequence is 1. And since 1 is greater than $\frac{4}{7}$ the fractions in the sequence must be increasing.

Therefore $\frac{5}{8}$ is greater than $\frac{4}{7}$.

We do this sort of thing frequently in our thinking, when for example we can predict where an argument is leading. We see that one action will lead to another, which leads to another and so on, with a predictable result.

EXAMPLE 3 - crushed triangle

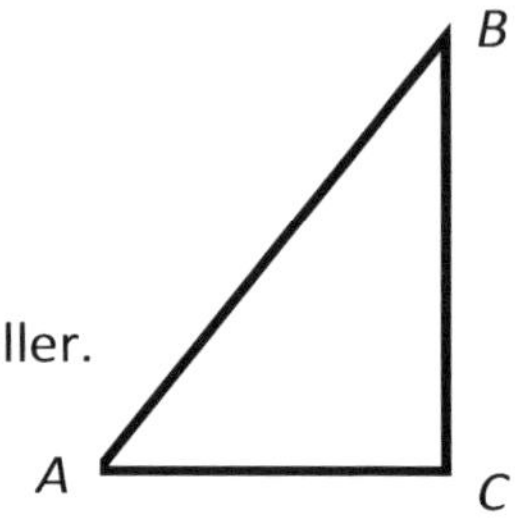

Suppose we have a triangle *ABC*, as shown.

Now suppose that the point *B* moves towards *C*.

As *B* gets lower the triangle gets smaller: its height gets smaller.

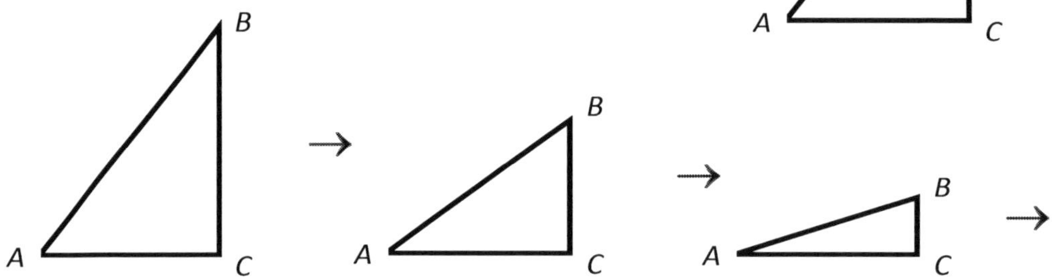

AB means the line that goes from *A* to *B*.

What is the limit of *AB* as *B* approaches and merges into *C*?

Imagine the shape changing as *B* gets closer and closer to *C* and you will see that as *B* merges with *C*, *AB* merges with *AC*.

So the **limit of *AB* is *AC***.

EXAMPLE 4

What would be the limit of ***BC*** as *B* goes to *C*?
What happens to the **line *BC*** as *B* goes to *C*?

The line *BC* becomes a point. We can say the limit is *C*.

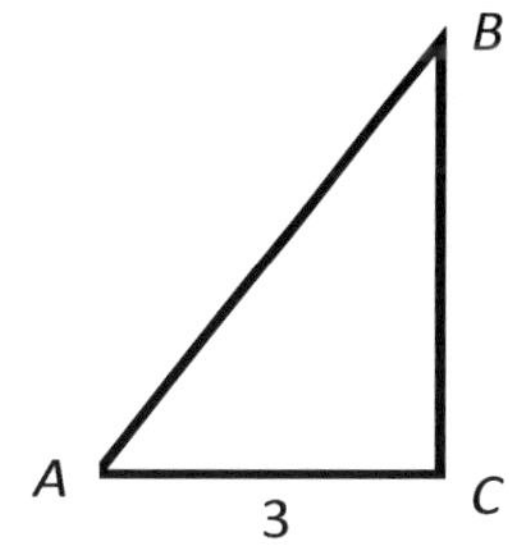

EXAMPLE 5 - limiting lengths

Now suppose that *AB* means the **length of the line *AB***.
So *AB* is a number, rather than a line.
And suppose the length of the base, *AC*, is 3 units.

What would you now say the limit of *AB* would be, as *B* goes to *C*?

Well, again as B approaches C, the length of *AB* will be closer and closer to the length of *AC*, which is 3. The **limit is 3**.

Exercise B (Answers Page 126)

What will be the limit of the length *BC*, as *B* approaches *C*?

As you will see later, these limits are extremely useful, so here we are just getting to know how to think about and imagine them.

EXAMPLE 6 - a secant

P and *Q* are two points on a circle.

A secant is a line that cuts a curve in two places, as you can see in the diagram below.

Now as *P* goes towards *Q* **around the circle**, what happens to the secant line as *P* approaches and merges into *Q*?

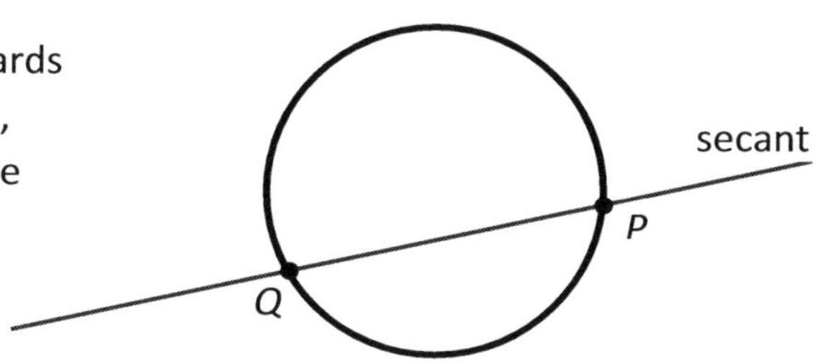

The line will pivot around *Q*.

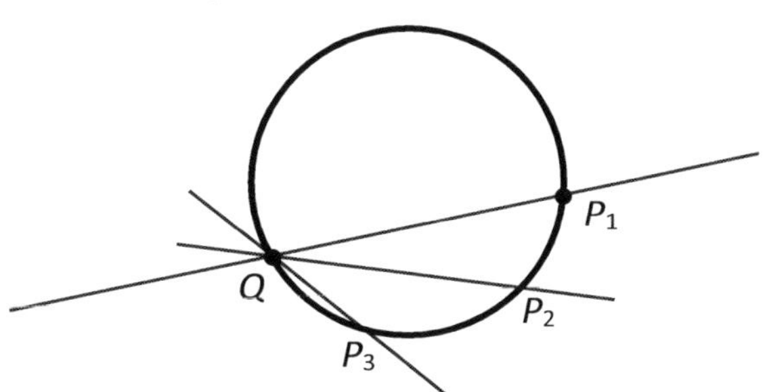

You may need to draw
some secant lines.

You should find that the
limit of the secant is a tangent:

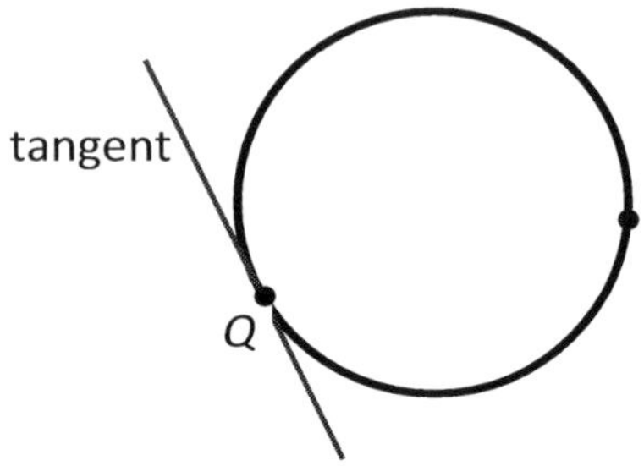

Next we apply this "secant-tangent" method to finding gradients of curves.

GRADIENT OF A CURVE

For a curve, the gradient is different at different points on the curve. If we want the gradient of the curve at a particular time or point, how can we get it?

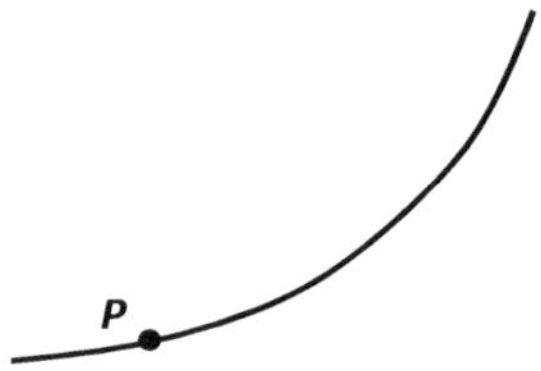

We can 'zoom in' on the point where we want the gradient.
If we zoom in enough, the line we see will look like a straight line. And we know how to find the gradient of a straight line.

If we zoom in, draw a straight line and then zoom out we will see that the straight line is actually a tangent to the curve at the point in question.

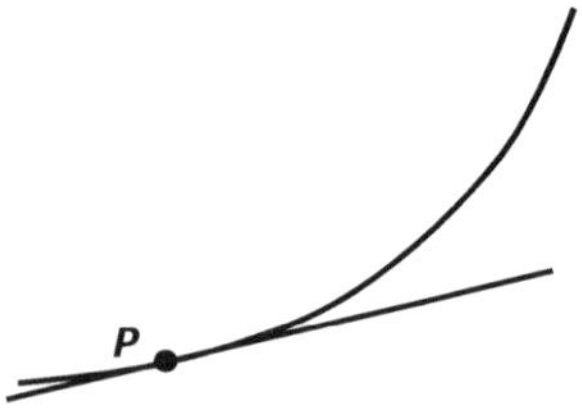

This brings us to the important conclusion that:

The gradient of a curve at a particular point is the gradient of the tangent at that point.

This can be illustrated as follows.

Suppose we have a curve as shown on the left below.
And suppose we want the gradient of the curve at the point *P*.

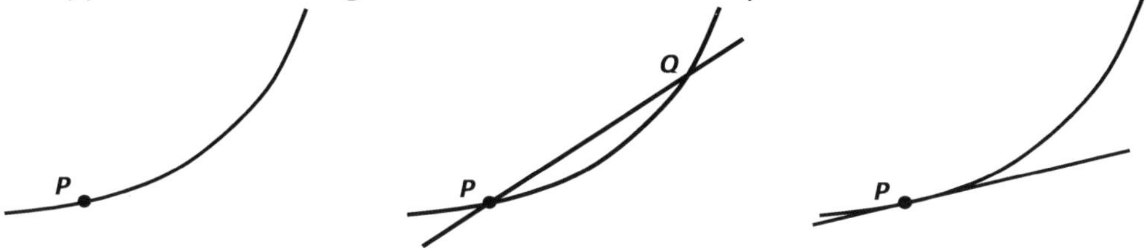

Draw a secant *PQ* as shown, and allow *Q* to move towards *P* along the curve.

What is the limit of the secant as *Q* merges with *P*?

As we have seen previously, it will be the tangent at *P*.
As *Q* approaches *P* the gradient of *PQ* approaches the gradient of the curve at *P*, and ultimately becomes the gradient of the tangent at *P*.

LAW OF GROWTH

The gradient of the tangent to a curve at a particular point describes its rate of change, which might be a rate of increase or a rate of decrease.

As a curve gets steeper the tangent gets steeper, so the gradient gets greater, all of which indicates that the rate of growth is greater.

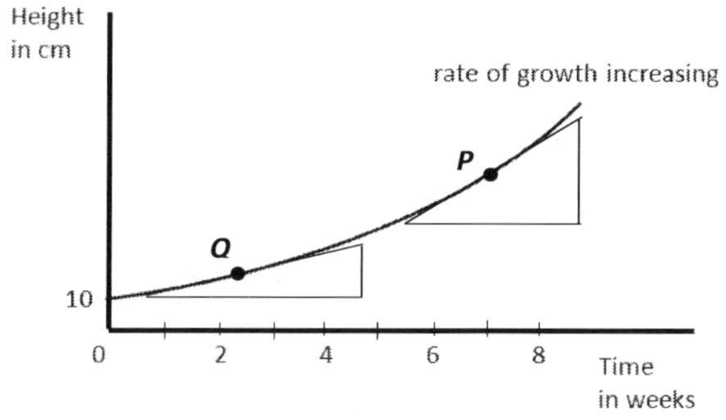

The rate of growth is greater at *P* than at *Q*, because the gradient of the tangent is greater.

So we have a **law of growth** - the curve above say, whose equation might be:

$$h = t^2 + 10.$$

And we have its gradient at each point, which describes its rate of growth.

So its **rate of growth** might be:

$$\text{rate of growth} = 2t,$$

meaning that at any time the tree is growing at a rate that is twice the time.

For example, after 3 weeks it will be growing at 6 cm per week.

Since we are modelling nature we cannot be certain that such predictions are correct, but they can still be very useful.

PRACTICE OF GRAPH WORK AND FINDING GRADIENTS

<u>**Exercise C**</u> (Answers Page 126)

a Copy and complete the table below for the graph of $y = x^2$.

x	-5	-4	-3	-2	-1	0	1	2	3	4	5
y	25					0	1	4	9		

b Take a sheet of graph paper and place it horizontally.
With the origin in the middle near the bottom of the sheet draw the:
x scale from -5 to 5 with 2 cm = 1 unit,
and the y scale from 0 to 25 with 2 cm = 5 units.

c Plot the 11 points carefully.

d Complete the table below and add 10 more points to your graph.

x	-4.5	-3.5	-2.5	-1.5	-0.5	0.5	1.5	2.5	3.5	4.5
y	20.25	12.25			0.25			6.25		20.25

e Draw the curve.

f Plot the points (5, 25) and (-1, 1) and find the gradient of the line joining them.

g Plot the points (-5, 25) and (-2, 4) and find the gradient of the line joining them.

Keep this graph for later.

Exercise D (Answers Page 126)

a Copy and complete the table below for the graph of $\mathbf{y = x^2 - 2x - 3}$.

x	-1	0	1	2	3	4
y	0	-3			0	5

b Place a sheet of graph paper horizontally, and with the origin near the centre draw the x scale from -1 to 4 with 2 cm = 1 unit, and the y scale from -4 to 5 with 1 cm = 1 unit.

c Plot the six points and draw the curve.

d By drawing tangents to the curve where $x = 3$, $x = 2$, $x = 1$, find the gradients of each of the three tangents.

Key Points of Chapter 1

From this chapter you should have got an idea how to think about limits by mentally following something until it reaches its final place.

You should have got a better understanding of the ways in which something may grow and how it is described, in terms of gradients.

AHEAD

Next we look at a simple way to get gradients of secants. Then in Chapter 3 we see how to combine this with limits to get gradients of curves.

CHAPTER 2

Gradient of a Secant

In this Chapter:

*** What is meant by gradient of a secant**
*** How to work it out**
*** And a simple way to get the answers more easily**

If we know the equation of a curve that describes some sort of growth, say $y = x^2$, we may need to know how to find the gradient at any point on it without having to draw the curve and measure the gradient of a tangent.

Here we start by seeing how to get the gradient of a secant, then in Chapter 3 we easily extend this to get the gradient of a tangent.

EXAMPLE 1

Suppose we have the curve given by the equation $y = x^2$ and we want to know the gradient of the secant joining the points P and Q on the curve where $x = 2$ and $x = 5$, as shown below.

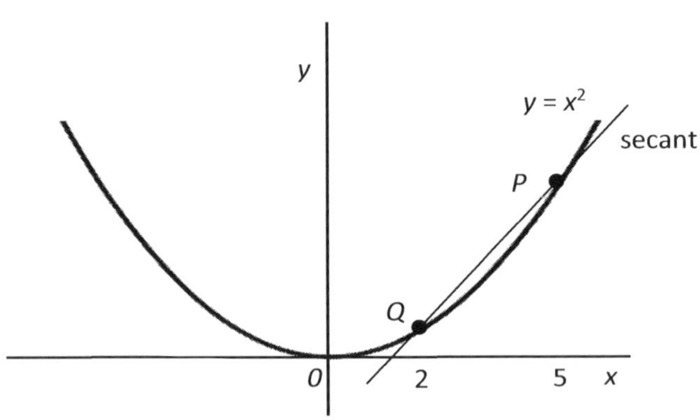

We can get the gradient of PQ by completing the triangle PQR shown and dividing the height of the triangle by its base.

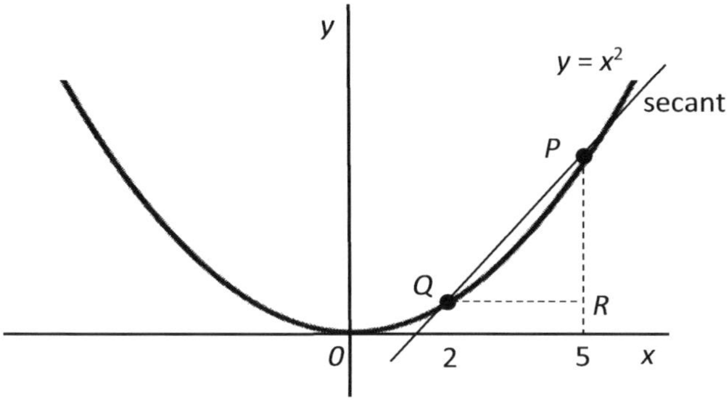

The coordinates of P are $(5, 5^2)$ or $(5, 25)$.
The coordinates of Q are $(2, 2^2)$ or $(2, 4)$.

So the gradient of $PQ = \dfrac{PR}{QR} = \dfrac{y\text{–coordinate of } P \text{ minus } y\text{–coordinate of } Q}{x\text{–coordinate of } P \text{ minus } x\text{–coordinate of } Q}$

Then since $P(5, 25)$ and $Q(2, 4)$ the gradient will be $\dfrac{25-4}{5-2} = \dfrac{21}{3} = 7$.

The secant gradient is **7**.

There is nothing at all difficult then in finding the gradient of a secant. We find the coordinates of the two points on the curve and divide the difference of the y-coordinates by the difference of the x-coordinates.

[This method will in fact give the secant gradient for any curve, not just for parabolas.]

There is something interesting to note here though which will give us the gradient more easily.

Exercise A (Answers Page 126)

For the curve $y = x^2$ find the gradient of the secant when P and Q have x-coordinates:

a 2 and 7 **b** 3 and 11 **c** -5 and 7 **d** -2 and -3

Do you notice anything?

There is a connection between the x-coordinates and the secant gradient.

You should have found that the gradient is the sum of the x-coordinates.

We can say:

> the gradient of a secant of $y = x^2$ joining points with x-coordinates p and q is $\boldsymbol{p + q}$.

This makes the work much easier.

Exercise B (Answers Page 126)
Use this result to find the gradient of the secant when P and Q have x-coordinates:

a 4 and 9 **b** 2 and 18 **c** -3 and 9 **d** -5 and -6

PROOF THAT THE SECANT GRADIENT IS $p + q$

In fact we can prove this result is true for all secants of $y = x^2$.

So we need to use some algebra.

Suppose p and q are the x-coordinates of P and Q.

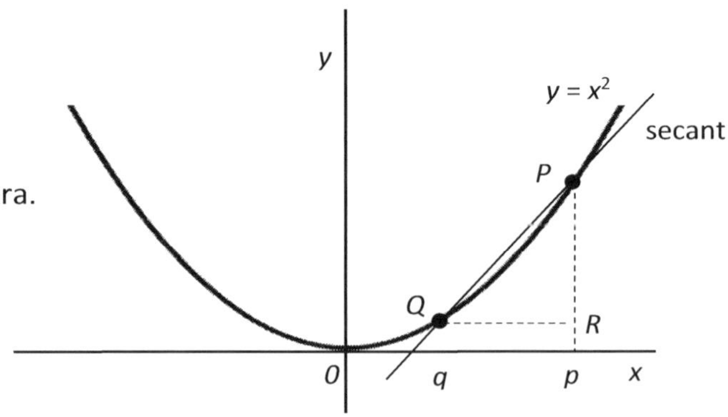

Then, since $y = x^2$, the coordinates of P and Q are:

$P(p, p^2)$ and $Q(q, q^2)$.

Now divide the difference of the y-coordinates by the difference of the x-coordinates:

$$\text{Gradient of } PQ = \frac{p^2 - q^2}{p - q}.$$

Now you may recall that $\boldsymbol{p^2 - q^2 = (p - q)(p + q)}$, so we can say:

$$\text{gradient of } PQ = \frac{(p-q)(p+q)}{p-q} = \boldsymbol{p + q}.$$

Which is what we wanted to prove.

This little formula will give us the gradient of any secant of $y = x^2$ given the x-coordinates of the two points on the curve.

We can write the gradient as $\frac{Dy}{Dx}$ because it's the:

Difference of the y-coordinates, Dy, divided by the Difference of the x-coordinates, Dx.

So:

$$\boxed{\frac{\boldsymbol{Dy}}{\boldsymbol{Dx}} = \boldsymbol{p + q}}$$

THE QUADRATICS $y = ax^2$

Well what about other parabolas? For example $y = 3x^2$.

The coefficient of x^2 is 3 here, rather than 1. (The coefficient is the number in front of the x^2.)

So given two x-values of points on the curve can we find the secant gradient?

Let's find the gradient of the secant for this curve for the line where $x = 7$ and $x = 2$.

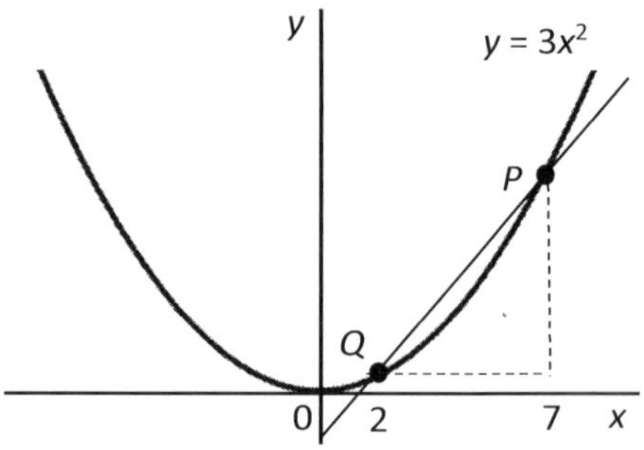

Again we need to find the y-coordinates of P and Q, and divide the difference of the y-coordinates by the difference of the x-coordinates.

We find: $P(7, 147)$ and $Q(2, 12)$.

So gradient of $PQ = \dfrac{Dy}{Dx} = \dfrac{147 - 12}{7 - 2} = \dfrac{135}{5} = 27$.

But we see that adding the given x-coordinates gives $2 + 7 = 9$, not 27, so the formula is not, 'add the x-coordinates' as it was before.

Exercise C (Answers Page 127)

a Copy and complete the following table:

Curve	x-coordinates	$\dfrac{Dy}{Dx} =$
$y = 3x^2$	$x = 2$ and $x = 7$	27
$y = 2x^2$	$x = 3$ and $x = 4$	
$y = 3x^2$	$x = 3$ and $x = 4$	
$y = 4x^2$	$x = 7$ and $x = 5$	
$y = 5x^2$	$x = 9$ and $x = 0$	
$y = -2x^2$	$x = 5$ and $x = 2$	
$y = 3x^2$	$x = 7$ and $x = -4$	

b Find the rule that gives the gradient from the x-coordinates.

c Adapt the proof on Page 14 to show that the gradient of the secant of $y = ax^2$ joining points with x-coordinates p and q is $\boldsymbol{a(p + q)}$?

d Confirm that this rule gives all the answers in column 3 of the table above.

Note that in $y = ax^2$ the a could be negative: meaning that the parabola is upside-down.

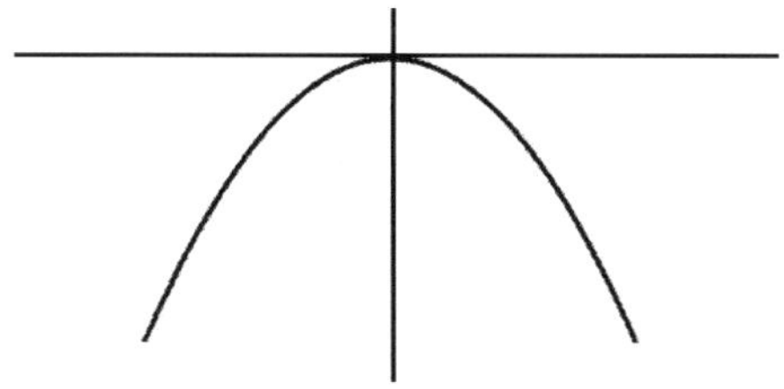

But our proof shows that the formula is still true because in the proof a was not restricted to being only positive.

The equation $y = ax^2$ represents all parabolas with their vertex at the origin.

But the family of curves represented by $y = ax^2$ is limited. To be able to find gradients for all parabolas we need to consider the quadratic equation $y = ax^2 + bx + c$ which represents all parabolas.

THE GENERAL QUADRATIC: $y = ax^2 + bx + c$

Let's start with a numerical example.

EXAMPLE 2

Suppose we have the parabola $y = 2x^2 + 3x + 4$, and we want to find the gradient of the secant between the points where $x = 5$ and $x = 1$.

We will find the y-coordinates of those points and divide the difference of the y-coordinates by the difference of the x-coordinates, just as before.

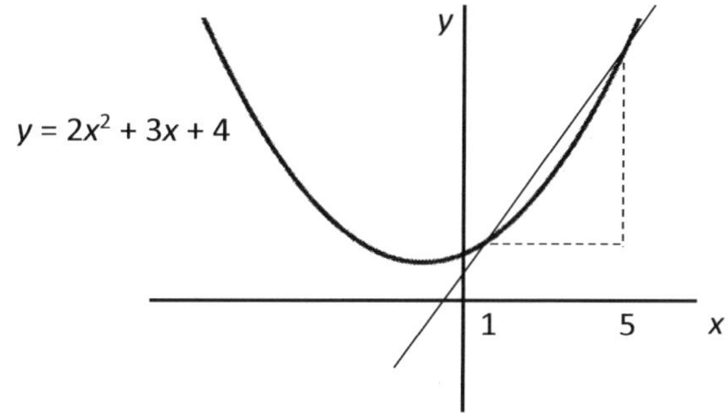

Then the points will be (5, 69) and (1, 9).

The secant gradient, $\dfrac{Dy}{Dx} = \dfrac{69-9}{5-1} = 15$.

Exercise D (Answers Page 127, 8)

a Find the secant gradients for the following and see if you can find the connection between the equation, the two x-coordinates, and the answer.

Curve, $y = ax^2 + bx + c$	***x*-coordinates,** p **and** q	$\frac{Dy}{Dx} =$	$a(p + q)$
$y = 2x^2 + 3x + 4$	$x = 5$ and $x = 1$	15	
$y = 2x^2 + 3x + 5$	$x = 4$ and $x = 1$		
$y = 3x^2 + x - 4$	$x = 3$ and $x = 1$		
$y = 4x^2 + 3x + 2$	$x = 5$ and $x = 2$		
$y = 5x^2 + 6x + 8$	$x = 3$ and $x = 0$		

Since the connection is not so obvious as before it will help if you also complete the last column above.

b Complete the formula: $\frac{Dy}{Dx} = a(p + q) + \ldots$

c Use this formula to complete the table below:

Curve, $y = ax^2 + bx + c$	***x*-coordinates,** p **and** q	$\frac{Dy}{Dx} = a(p + q) + b$
$y = 3x^2 + 4x + 1$	$x = 4$ and $x = 7$	
$y = -2x^2 + 3x - 4$	$x = 5$ and $x = 3$	
$y = 4x^2 + 5$	$x = 4$ and $x = 1$	
$y = x^2 - 2x + 15$	$x = 9$ and $x = -8$	

d Why do you think c does not appear in the formula?

e Prove the result in **b** above with the method used on Page 14.

The secant gradient for any parabola is given by: $\boldsymbol{\frac{Dy}{Dx} = a(p + q) + b.}$

Note that our previous results are contained within this formula.

1) For example, we saw that for $y = x^2$, $\frac{Dy}{Dx} = p + q$.

   But comparing $y = x^2$ with $y = ax^2 + bx + c$ we see that they are the same when $a = 1$, $b = 0$ and $c = 0$.

   And if we put these values into $\frac{Dy}{Dx} = a(p + q) + b$ we get $p + q$.

2) Similarly we saw that for $y = ax^2$, $\frac{Dy}{Dx} = a(p + q)$.

   But comparing $y = ax^2$ with $y = ax^2 + bx + c$ we see that they are the same when $b = 0$ and $c = 0$.

   And if we put these values into $\frac{Dy}{Dx} = a(p + q) + b$ we get $a(p + q)$.

Key Points of Chapter 2

The gradient of a secant of a parabola $y = ax^2 + bx + c$, which passes through points with x-coordinates p and q, is easy to find using the simple formula:

$$\boldsymbol{\frac{Dy}{Dx} = a(p + q) + b.}$$

AHEAD

In the next chapter we build on this to get gradients of tangents by using limits.

CHAPTER 3

Gradient of a Tangent

In this Chapter:

*** How we can use the result of the last chapter, together with limits, to get a simple formula that will give us the gradient at any point on any parabola.**

Now let's go back to the parabola $y = x^2$.

We saw that the gradient of the secant joining the points where $x = p$ and $x = q$ can be found by simply adding the x-coordinates: $\frac{Dy}{Dx} = p + q$.

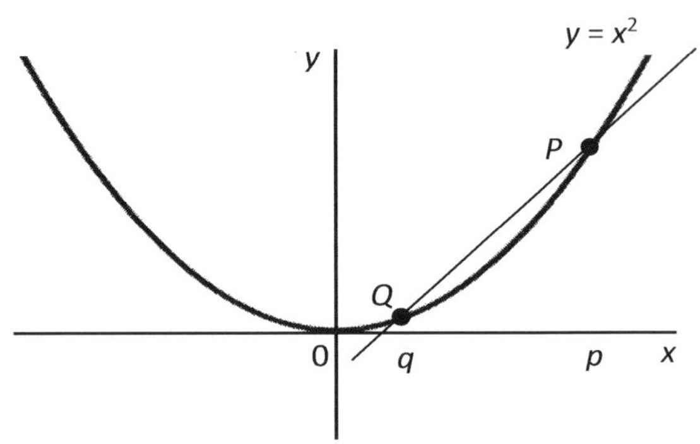

USING LIMITS

Imagine the point Q moving along the curve towards P.

What will happen to the **secant** as Q approaches and merges with P?

It will become the tangent at P.

And as Q merges with P, q becomes equal to p.

So what is the gradient of the tangent at P?

Instead of $p + q$, for the gradient of the secant, it will be $p + p$, which is $2p$, or $2x$ since p is an x-coordinate. [Please see the note on Page 115]

We can summarise this result as follows:

The gradient of the tangent at any point on the curve $y = x^2$ is twice its x-coordinate.

Or:

The gradient at any point on the curve $y = x^2$ is $2x$.

Exercise A (Answers Page 129)

a What is the gradient at the point on $y = x^2$ where $x = 2$?

b What is the gradient at the point where $x = 3$?

c What is the gradient at the point (4, 16)?

d Copy and complete the table below for the graph of $\mathbf{y = x^2}$.
If you have the graph from Exercise C, Chapter 1 you can use that here. In which case you can move on to question **i**.

x	-5	-4	-3	-2	-1	0	1	2	3	4	5
y	25					0	1	4	9		

e Take a sheet of graph paper and place it horizontally.
With the origin in the middle near the bottom of the sheet draw the:
x scale from -5 to 5 with 2 cm = 1 unit,
and the y scale from 0 to 25 with 2 cm = 5 units.

f Plot the 11 points carefully.

g Complete the table below and add 10 more points to your graph.

x	-4.5	-3.5	-2.5	-1.5	-0.5	0.5	1.5	2.5	3.5	4.5
y	20.25	12.25			0.25			6.25		20.25

h Draw the curve.

i Check the results in **a, b, c** above by drawing tangents and checking that the gradient is twice the *x*-coordinate.

Test for a few other points until you are satisfied.

j Is there any point you could choose on the curve where the gradient is not twice the *x*-coordinate? What about points to the left of the origin?

GRADIENT OF ANY PARABOLA

We know that the secant gradient for the general parabola, represented by $y = ax^2 + bx + c$, can be found using:

$$\frac{Dy}{Dx} = a(p + q) + b.$$

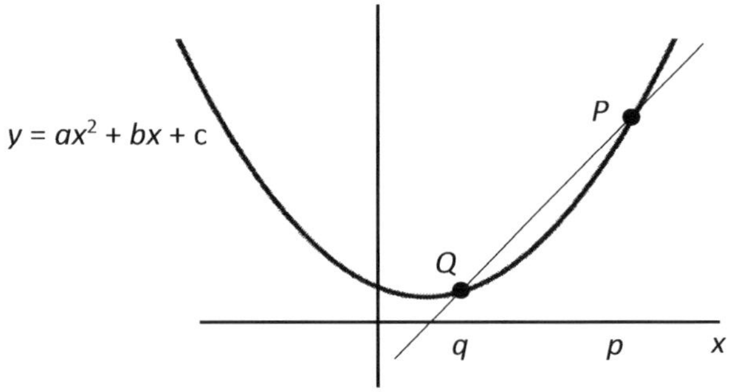

We allow Q to approach P, as before, and put $q = p$:

So $a(p + q) + b$ becomes $a(p + p) + b = 2ap + b$.

Tangent gradient = $2ap + b$.

Or, tangent gradient = $2ax + b$, since p is an x-coordinate.

For $y = ax^2 + bx + c$ the tangent gradient = $2ax + b$.

EXAMPLE 1

Find the gradient of $y = 2x^2 + x + 7$ where $x = 3$.

- Here we see $a = 2$, $b = 1$.

  Therefore the gradient = $2ax + b = 2 \times 2 \times 3 + 1 =$ **12**.

Note that the 7 has no effect on the answer.

Exercise B (Answers Page 129) Find the gradient of the curve at the given point.

a $y = 3x^2 + 4x + 9$ where $x = 5$

b $y = x^2 - 3x + 4$ where $x = 1$

c $y = 2x^2 + 8x + 90$ where $x = 2$

d $y = x^2 + 5x - 1$ where $x = 3$

e $y = 2x^2 + x$ where $x = 3$

f $y = x^2 - 4$ where $x = -4$

THE QUICK WAY TO GET THE GRADIENT FORMULA

Note carefully the quick way to get the gradient formula for $\mathbf{3x^2 + 5x + 7}$:
we look at $\mathbf{3x^2}$ and multiply the power (2) by the coefficient (3) to get the 6, which we put down as $\mathbf{6x}$,
Then we append the coefficient of x in $\mathbf{3x^2 + 5x + 7}$, which is **5**, to get $\mathbf{6x + 5}$ as the answer.

This is shown in the diagram below.

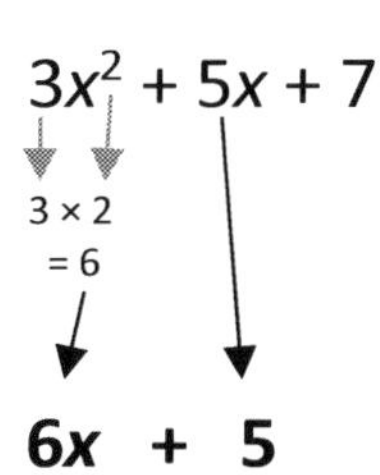

This gives the gradient formula, like $\mathbf{6x + 5}$, for any quadratic more easily.

EXAMPLE 2

Find the gradient of the tangent to $y = -2x^2 + 1$ at the point on the curve where $x = 3$.

- We have gradient = $-4x$.

  So at $x = 3$, gradient = -12.

<u>**Exercise C**</u> (Answers Page 129) Find the tangent gradient:

a at $x = 2$ on the curve $y = x^2 + 4x + 1$

b at $x = -3$ on the curve $y = 2x^2 - x - 5$

c at $x = 0$ on the curve $y = 5 - 3x^2$

d at the point (4, 42) on the curve $y = x^2 + 7x - 2$

e If $y = 2x^2 - 3x + 4$ find the gradient at $x = 5$.

f If $y = 3x^2 + 3x - 4$ find the gradient at $x = 5$.

THE SPECIAL SYMBOL $\frac{dy}{dx}$

In fact we have a special symbol for gradients: $\frac{dy}{dx}$.

So, for the quadratic $y = ax^2 + bx + c$, we write $\frac{dy}{dx}$ = $\mathbf{2ax + b}$.

You may ask, why do we have $\frac{dy}{dx}$ instead of $\frac{Dy}{Dx}$? And the answer is that Dy and Dx are y and x-differences. And if we allow Q to approach and merge in P these differences become infinitely small, so that $\frac{dy}{dx}$ represents the value that the ratio of two quantities is approaching. This is a different meaning to $\frac{Dy}{Dx}$ and therefore needs a different symbol.

$\frac{dy}{dx}$ is standard notation for the gradient of a curve.

> For the quadratic $y = ax^2 + bx + c$
>
> $$\frac{dy}{dx} = 2ax + b$$

EXAMPLE 3

Given $y = -4x^2 + 17x - 1$ find $\frac{dy}{dx}$.

- We multiply 2 by -4 to get -8.

  So $\frac{dy}{dx} = -8x + 17$.

EXAMPLE 4

Given $y = 5 - 3x^2$ find the gradient formula.

- Here the coefficient of x is zero (which means that $b = 0$), so we just multiply 2 by -3 and give:

  $\frac{dy}{dx} = -6x$.

EXAMPLE 5

If $y = mx^2 + nx + r$ find the gradient formula.

- We need to apply $\frac{dy}{dx} = 2ax + b$ where $a = m$ and $b = n$.

  We get $\frac{dy}{dx} = 2mx + n$.

Exercise D (Answers Page 129) Find $\frac{dy}{dx}$ for:

a $y = 3x^2 + 4x + 5$ **b** $y = -7x^2 + x + 5$ **c** $y = 8 - 9x + 3x^2$

d $y = 9x^2 + 16x + 100$ **e** $y = px^2 + qx + s$ **f** $y = ax^2 + 4x + 5$

g $y = bx^2 + 4x + 5$ **h** $y = bx^2 + cx + d$

i If $y = 2x^2 - 3x + 4$ find $\frac{dy}{dx}$ at $x = 0$

j If $y = x^2 - 2x - 3$ find $\frac{dy}{dx}$ at $x = 3$, 2 and 1.

k Go back to the gradients you found by measurement in Exercise D of Chapter 1. Check that the gradients you got in **j** above are about the same.

FINDING THE POINT WHICH HAS A GIVEN GRADIENT

We know how to find the gradient at any point on a parabola. Let's see how we can do the opposite: find a point that has a given gradient.

EXAMPLE 6

For example, at what point on the parabola $y = x^2 + x - 4$ is the gradient equal to 3?

You could draw the parabola and look for where the gradient is 3. Or we can find it accurately as follows.

- Since we know the gradient is 3 we can obtain the gradient formula for this parabola and equate that to 3.

And since $\frac{dy}{dx}$ = 2x + 1, we have **2*x* + 1 = 3**.

Solving this equation we get $x = 1$.

To get the y-coordinate corresponding to $x = 1$ on the curve we can substitute $x = 1$ in $y = x^2 + x - 4$.

We get $y = -2$ and so the required point is **(1, -2)**.

Exercise E (Answers Page 130) Find the coordinates of the point on the parabola which has the gradient given.

a $y = x^2 + x - 4$, $\frac{dy}{dx} = 5$

b $y = 2x^2 + 3x + 4$, $\frac{dy}{dx} = 3$

c $y = 18 - x^2$, $\frac{dy}{dx} = 1$

d $y = x^2 + x - 4$, $\frac{dy}{dx} = 0$

e $y = x^2 + 4x + 5$, $\frac{dy}{dx} = -8$

f $y = 5x^2$, $\frac{dy}{dx} = 20$

DIFFERENTIATION

The process of obtaining the gradient formula is called 'differentiation'.

EXAMPLE 7

So, for example, asked to differentiate $y = 5x^2 - 8x + 9$, we would give the answer:

- $\frac{dy}{dx} = 10x - 8$.

In fact 'differentiation' is one of the two branches of Calculus. The other branch is called 'Integration' and we come to that later.

EXAMPLE 8

Differentiate C where $\boldsymbol{C = q^2 - 5q + 70}$.

- Here we have C and q instead of y and x, so we apply the same procedure but use C and q.

$$\frac{dC}{dq} = 2q - 5.$$

Note we also have $\frac{dC}{dq}$ instead of $\frac{dy}{dx}$.

Exercise F (Answers Page 130) Differentiate the following (Include the correct $\frac{dy}{dx}$ symbol as in Example 8):

a $y = x^2 + 41x + 52$ **b** $y = 2x^2 - x - 5$ **c** $y = 6 - \frac{1}{2}x^2$

d $y = 3t^2 + t - 7$ **e** $N = 3y^2 + 9$ **f** $x = 15 - 5t^2$

g $y = x^2 + x - \pi$ **h** $T = x^2 + \frac{2}{3}x + 3$ **i** $A = \pi r^2$

Just to remind you: what we get when we differentiate is the gradient of the curve at any point, which is also the rate of growth.

THE OPERATOR

There is alternative way to indicate differentiation that is often used, and this can be more convenient because it involves only x and not y.

EXAMPLE 9

For example we can write $\frac{d}{dx}(5x^2 - 8x + 9) = 10x - 8$.

The symbol $\frac{d}{dx}$ tells you that what follows it is to be differentiated. It is called an operator.

It tells you to carry out the differentiation operation on the expression that follows.

EXAMPLE 10

So, similarly, $\frac{d}{dx}(7x^2 + 6x - 55) = 14x + 6$.

Exercise G (Answers Page 130) Find:

a $\frac{d}{dx}(4x^2 - x - 1)$ **b** $\frac{d}{dx}(x^2 + 7)$ **c** $\frac{d}{dx}(-5x - 4x^2)$

d $\frac{d}{dx}(20 - x^2)$ **e** $\frac{d}{dx}(x^2 + \pi x)$ **f** $\frac{d}{dx}(2x^2 + 5.6)$

Key Points of Chapter 3

* We arrive at the simple but very important result that the gradient at any point on a parabola $y = ax^2 + bx + c$ can be found from $\frac{dy}{dx} = 2ax + b$.

* The process of getting $\frac{dy}{dx}$ is called "differentiation".

* An alternative symbol can be used. For example $\frac{d}{dx}(4x^2 - 7x) = 8x - 7$.

AHEAD

We will extend this method in Chapter 5: to cubics, quartics etc. But first (in Chapter 4) we will see some actual applications of the method of this chapter in finding maximum and minimum values.

CHAPTER 4

Optimisation

In this Chapter:

*** How to find maximum and minimum values using calculus**

*** How to use calculus to help sketch curves**

*** Some practical examples**

*** How to form your own equation and optimise**

INTRODUCTION

Optimisation is the process of finding a maximum or minimum value. A businessman for example will want to maximise his profit, so he would want to know the best price to sell his items at to maximise the profit. He will also want to maximise the number of items he sells and also minimise costs.

We may want to maximise the water supply in a pipe, maximise the strength of a metal rod used in an aircraft construction. We may want to minimise the load on a wall or pollutants in the atmosphere. And it's calculus that supplies the answers.

The most significant feature of a parabola is its turning point, or vertex. And in fact it is very easy to get the coordinates of the turning point.

What will be the gradient at the vertex of a parabola?

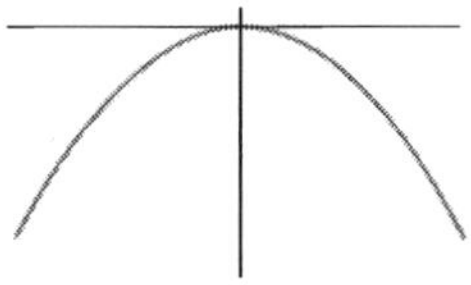

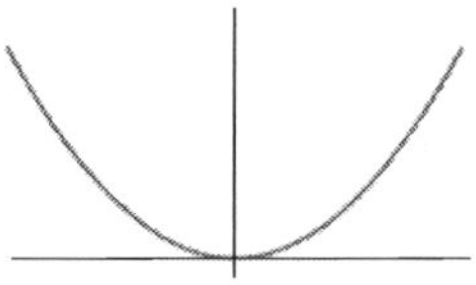

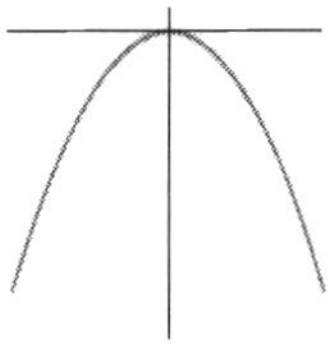

It will always be zero.

So given the equation of any parabola we can find its maximum or minimum by **differentiating it and equating the result to zero**.

At the vertex of a parabola $\frac{dy}{dx} = 0$.

And knowing the location of a turning point makes sketching graphs easier too.

SKETCHING PARABOLAS

EXAMPLE 1

What are the coordinates of the vertex of the parabola $y = x^2 - 4x + 7$.

- We differentiate and equate to zero: $2x - 4 = 0$.

  The solution to this is $x = 2$, so the x-coordinate of the turning point is 2.

  Substituting $x = 2$ in $y = x^2 - 4x + 7$ we get $y = 3$.

  So the coordinates of the turning point are (2, 3).

This helps us to sketch the curve.
We know that in this example the coefficient of x^2 is positive,
which means the parabola is U-shaped.
So we can make a rough sketch.

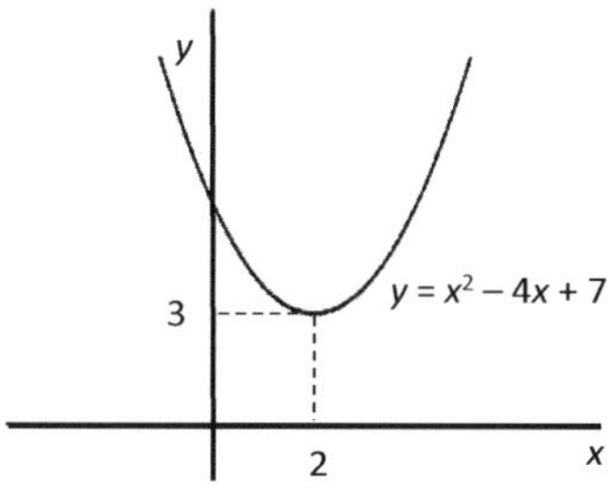

Also, the intercept of the curve on the y-axis is easy to see because on the y-axis $x = 0$.

So by putting $x = 0$ in $y = x^2 - 4x + 7$ we see that $y = 7$ and we could add this information to our graph, as you can see here.

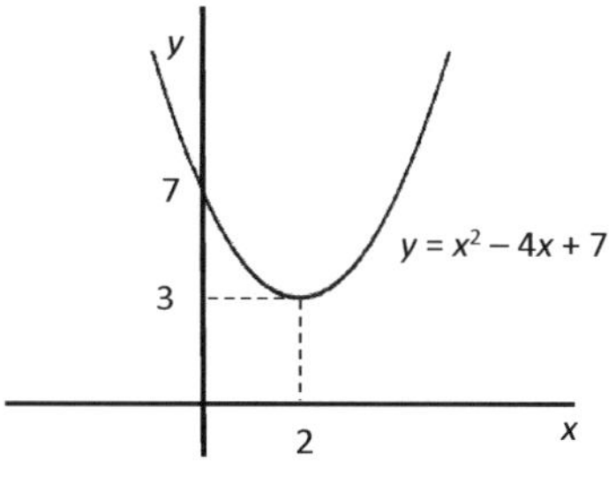

Since the graph is a parabola there will be a minimum value for y, which is 3.
y can take the value 3 and any value greater than that, but it can't be less than 3, so we say the minimum value is 3.

To find the value of x for the maximum or minimum point: differentiate, equate to zero and solve.

EXAMPLE 2

Sketch the graph of $y = 3 - 12x - 2x^2$.
Include the coordinates of the vertex and the intercept value on the y-axis.

- We note that the coefficient of x^2 is negative so the parabola is upside-down.

  We differentiate and equate to zero:
  $-12 - 4x = 0$.

  This gives $x = -3$, and substituting this we get $y = 3 + 36 - 18 = 21$.

  For the intercept on the y-axis we put $x = 0$ and get $y = 3$.

  So the graph is as shown.

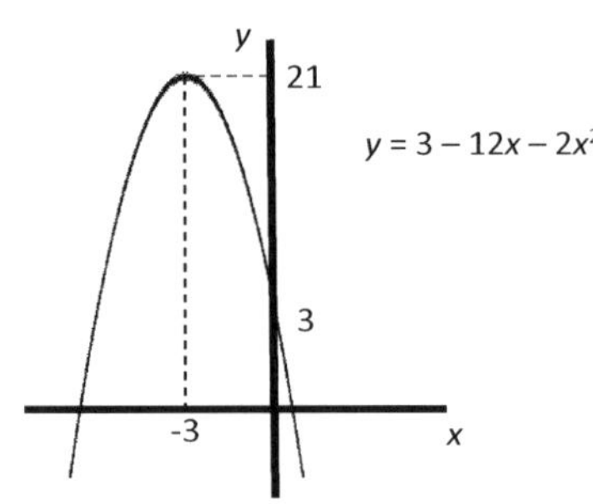

Exercise A (Answers Page 130, 1)

Sketch the following curves.
Include the coordinates of the vertex and the intercept value on the y-axis.

a $y = x^2 - 14x + 60$ **b** $y = 3 + 2x - x^2$ **c** $y = -x^2 + 8x - 20$

d $y = 2x^2 + 8x + 1$ **e** $y = 3x^2 - 6x$ **f** $y = x^2 - 6x + 9$

OPTIMISING

EXAMPLE 3

What is the maximum value of $y = 3 + 20x - 2x^2$?

- We don't need to sketch this curve. We can see that it's an upside-down parabola (because the coefficient of x^2 is negative) and so it makes sense to talk about a maximum value.

  That value will simply be the *y*-coordinate of the vertex.
  And we can get that by first finding the *x*-coordinate of the vertex using calculus.

  $\frac{dy}{dx} = 20 - 4x = 0.$
  Therefore $x = 5$, and so $y = 3 + 20x - 2x^2 = 3 + 100 - 50 = 53$.

  The maximum value of $y = 3 + 20x - 2x^2$ is **53**.

EXAMPLE 4

Find the minimum value of $7 + 6x + x^2$.

- We have no '*y*' here. But we can still differentiate and equate to zero:

  $6 + 2x = 0$ and so $x = -3$.

  Putting $x = -3$ into the given quadratic we get $7 + 6(-3) + (-3)^2 = -2$.

  The minimum value of $7 + 6x + x^2$ is **-2**.

Exercise B (Answers Page 131)

a Find the maximum value of $y = 3 - 2x - x^2$

b Find the minimum value of $y = 2x^2 + 12x - 5$

c Find the minimum value of $y = 3x^2 - 12x + 1$

d Find the maximum value of $y = 18x - 3x^2$

PRACTICAL EXAMPLES

EXAMPLE 5

A projectile is fired and flies in a parabolic path whose height *h* at any time *t* is given by the equation:

$h = 200t - 5t^2$, where the units are meters and seconds.

Find the maximum height reached by the projectile and the time it takes to reach its greatest height.

- We note that when $t = 0$, $h = 0$, so the path will look something like this:

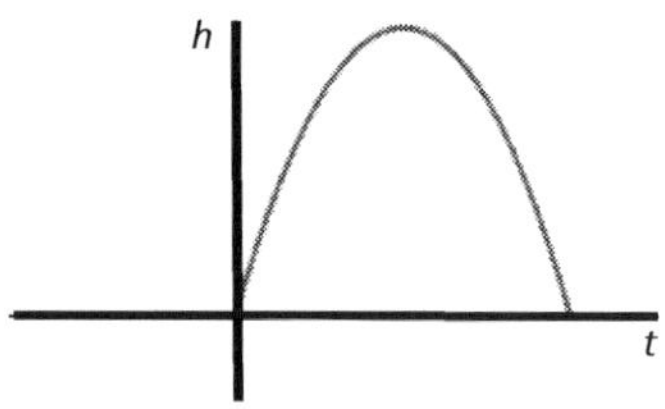

At the maximum height the gradient of the curve will be zero.

So $200 - 10t = 0$
This gives $\boldsymbol{t}$ **= 20 seconds** (it takes 20 seconds to reach its maximum height).

- The maximum height is then $200t - 5t^2 = 200\times20 - 5\times20^2 =$ **2000 m**.

EXAMPLE 6

The cost of producing *n* items in a factory is given by the equation:

$\boldsymbol{C = n^2 - 50n + 700}$, where *C* is the cost in $ of producing *n* items.

How many items should be produced if costs are to be minimised, and what is the minimum cost?

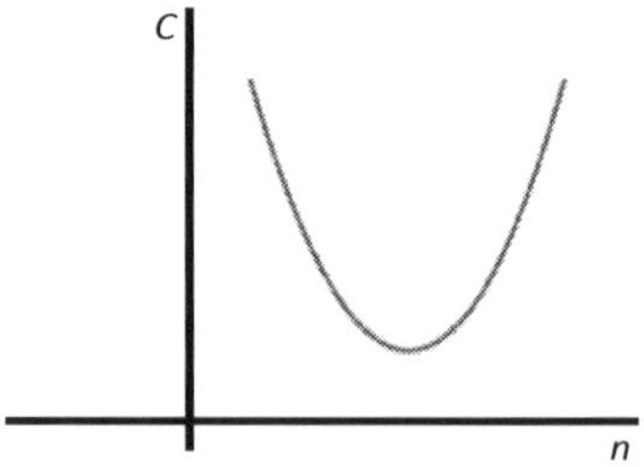

- We have *C* and *n* here instead of *y* and *x*.

  We see we have a parabolic shape like the one shown and we need the value of *n* at the lowest point.

- So as before we differentiate and equate to zero. We get $2n - 50 = 0$.

  And so $n = 25$.

That means, if we produce 25 items the cost will be minimised.

- To get that minimum cost we put $n = 25$ in the quadratic:

  $\boldsymbol{C = n^2 - 50n + 700} = 25^2 - 50 \times 25 + 700 = \mathbf{75}$.

  Since we were told that the cost is in dollars we can say that the minimum cost is **$75**.

EXAMPLE 7

Suppose you have this formula for revenue (income) in terms of price:

$$\boldsymbol{R = 10x - 2x^2}$$

where R is revenue in 1000s of dollars, and x is the price in dollars.

What price will maximise revenue?
And what is the maximum revenue?

- We differentiate $\boldsymbol{10x - 2x^2}$ to get $\boldsymbol{10 - 4x}$.

  Equate to zero and solve: $\boldsymbol{10 - 4x = 0}$ which gives $\boldsymbol{x = 2.5}$.

  So the price that maximises revenue will be **$2.50**.

- For the revenue we put $\boldsymbol{x = 2.5}$ in $\boldsymbol{R = 10x - 2x^2}$.

  Then $R = 10\text{x}2.5 - 2\text{x}2.5^2 = 25 - 12.5 = \mathbf{12.5}$.

  Since R is in 1000s of \$, the maximum revenue is **$12,500**.

Exercise C (Answers Page 131)

a A projectile is fired and flies in a parabolic path. Its height at any time t is given by the equation:

$h = 18t - t^2$, where the units are meters and seconds.

Find the maximum height reached by the projectile and the time it takes to reach its maximum height.

b If the cost in dollars of producing n items is given by $C = 2n^2 - 200n + 5120$ find the value of n that minimises C, and find the minimum cost.

c The equation $P = -x^2 + 20x - 50$, gives the profit, P, obtained when producing some item, in thousands of dollars, and x is the price of the item in dollars.

What price will give the maximum profit, and what will the maximum profit be?

FORMING YOUR OWN EQUATION

Sometimes we need to formulate our own equation.

Remember, we are involved with two variables here: the thing we are trying to maximise or minimise, say h, and some other variable that h depends on, say t.

So, for example, for the projectile earlier we had $h = 18t - t^2$.

h was the height and we wanted to find the maximum value of h, and t was time, which the height depended on.

If we are to construct our own equation we need to be sure what the two variables are that we are going to choose. The first one will be obvious: you give a letter to the thing you are trying to maximise or minimise. The other variable you need to think about and sometimes there is more than one choice.

EXAMPLE 8

Farmer Phil wishes to fence off a rectangular enclosure against a wall using 600 m of fencing, so that his cows can dine on the nice green grass. Find the dimensions of the rectangle so that the area is maximised.

- Clearly there are many rectangular shapes that Farmer Phil could have with his 600 m of fencing. And the area of the enclosure will vary.

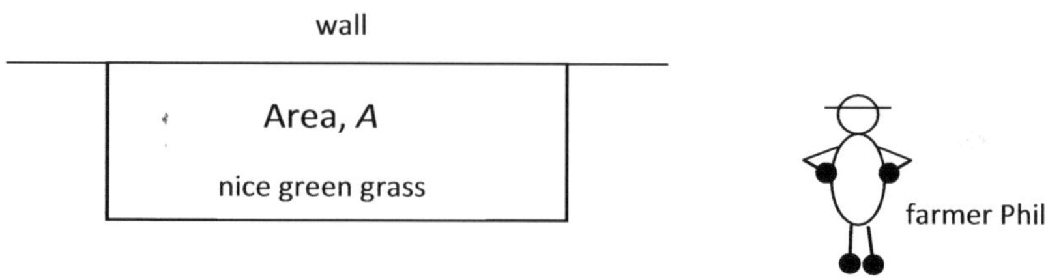

First we draw a diagram like the one above (the inclusion of Farmer Phil is optional).

And since we want to maximise the area we give that a letter, say A.

Now since we are asked for the dimensions of the rectangle (its base and height), we could choose one of those for the other variable and call it x as shown below.

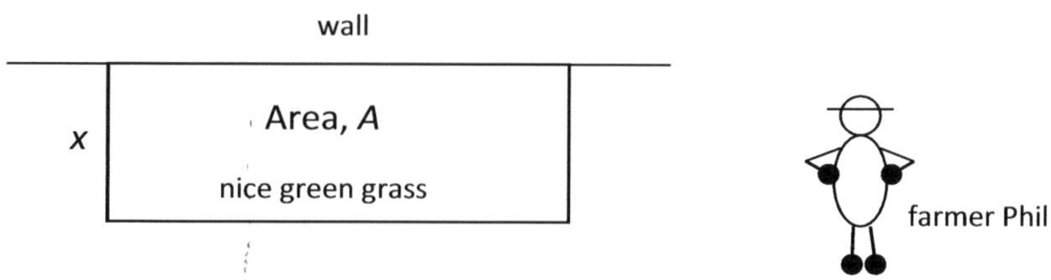

So x could be 100 m for example, and the base of the rectangle would then be 400 m, and the area would be $400 \times 100 = 40{,}000$ m^2. Make sure you agree with this.

But is that the maximum area?

We can find out by using calculus. First we need to set up an equation: we want to get an equation that includes both *A* and *x*.

Now the area of a rectangle is base × height, and we know the height is *x*.

Since we have 600 m of fencing we see that we have to take twice the height from the 600 to get the base.

And that means the base must be $600 - 2x$.

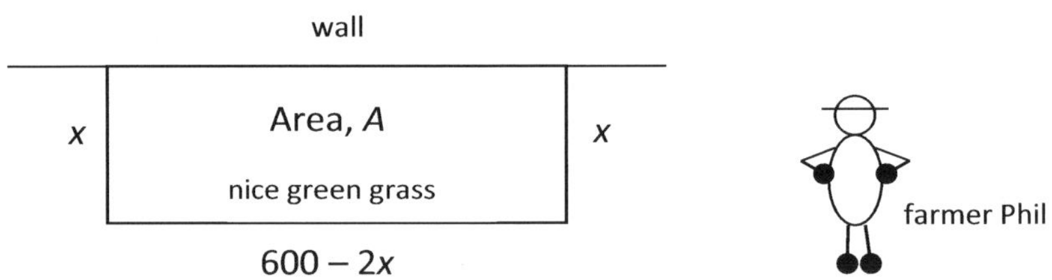

Now we can set up our equation: *A* = base × height, so

$A = (600 - 2x)x$.

To maximise *A* now we will differentiate and equate to zero.

But first we had better multiply the bracket out:

$A = 600x - 2x^2$.

We can now say $\frac{dA}{dx} = 600 - 4x = 0$.

And solving: $x = 150$.

The problem is now solved: the height of the rectangle must be **150 m** and the base will be $600 - 2x$ = **300 m**.

If you would like to check this you can work out some areas for different values of *x*. You will find that you cannot find a value of *x* that gives a greater area than the one we have found.

<u>Exercise D</u> (Answers Page 131)

Farmer Phil has a daughter, Daisy, who wants to copy Dad's idea by making a rectangular enclosure against a wall for her pet rabbit. She has 16 feet of fencing. Form an equation and find out what dimensions she should have to maximise the area?

Key Points of Chapter 4

* We can find turning points of parabolas by differentiating, equating the result to zero and solving.

* This helps us to sketch graphs.

* It also enables us to find maximum and minimum values which has many practical uses.

AHEAD

Next we extend the method from Chapter 3 and see how it can be applied to many other curves. In Chapters 6 and 7 we extend further to include powers that are fractions or negative.

CHAPTER 5

Cubics and Beyond

In this Chapter:

*** We extend the method introduced in Chapter 3 to cubic and higher order equations.**
*** We are introduced to the second derivative**

In Chapter 3 we saw that we could easily convert the formula for the gradient of a **secant** of $y = ax^2 + bx + c$ to give us the gradient of a **tangent**.

Here we will be using the same method to find gradient formulae for cubics, quartics etc.

So by the time you get to the end of the chapter you will be able to immediately give the gradient formula for polynomials like:

$$y = 5x^3$$

$$y = 2x^5 + 4x^4 - 8x^3 + x^2 - 9$$

In Chapter 2 we used the well-known formula: $\mathbf{p^2 - q^2 = (p - q)(p + q)}$, for the difference of two squares. We found the gradient of a secant of a parabola with it.
[This was $\boldsymbol{a(p + q) + b}$]

And in Chapter 3 we put $q = p$ and obtained the gradient of a tangent.
[This was $\boldsymbol{2ax + b}$]

Now, using further formulae, for the difference of two cubes, difference of two fourth powers etc. we can get gradients of other curves too.

We will do this in three stages:

[1] First we'll look at $y = x^3$, x^4, x^5 etc.
[2] Then $y = ax^3$, ax^4, ax^5 etc.
[3] Then combinations of these like $y = ax^3 + bx^2 + cx + d$.

DIFFERENCE OF SQUARES, CUBES ETC.

The formulae for finding the difference of two squares, difference of two cubes etc. are listed below.

$$\boldsymbol{p^2 - q^2 = (p - q)(p + q)} \qquad (1)$$

$$\boldsymbol{p^3 - q^3 = (p - q)(p^2 + pq + q^2)} \qquad (2)$$

$$\boldsymbol{p^4 - q^4 = (p - q)(p^3 + p^2q + pq^2 + q^3)} \qquad (3)$$

etc.

These formulae will be used in our gradient work and will come up again when we are finding areas under curves later on. The first formula is very well-known and very useful.

<u>**Exercise A**</u> (Answers Page 131, 2)

a Confirm each of these formulae is correct by multiplying the two factors on the right-hand side.

b Show that formula (2) works for $5^3 - 3^3$ by showing that the value of $5^3 - 3^3$ is obtained by subtraction and also by substituting into the right-hand side of the formula.

c By following the patterns in the above list of three formulae write down the next formula: the one for $p^5 - q^5$.

d What is the connection between the power on the left-hand side of each formula and the number of terms in the second bracket?

e What can you say about the power on the very first term in each formula and the power on the first term in the second bracket?

[1] GRADIENT OF $y = x^3, x^4, x^5$ etc.

In finding **a secant gradient for $y = x^3$**, we need to find $\frac{p^3 - q^3}{p - q}$, that is the difference of the y-coordinates divided by the difference of the x-coordinates (see diagram below).

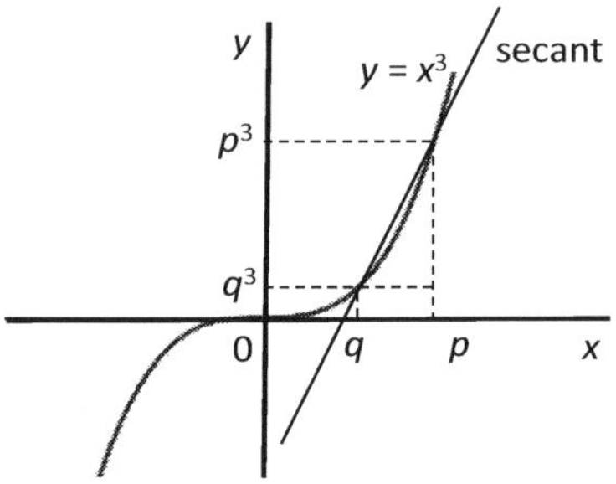

And $\frac{p^3 - q^3}{p - q}$ is available from formula (2) above: it is $(p^2 + pq + q^2)$.

Then putting $q = p$ (as we did before for the parabola) we get the tangent gradient as $3p^2$ or $3x^2$, since p is an x-coordinate.

So we can say that if $y = x^3$ then $\frac{dy}{dx} = 3x^2$.

Similarly for $y = x^4$, we need $\frac{p^4 - q^4}{p - q}$ which is $(p^3 + p^2q + pq^2 + q^3)$, by formula (3) above.

Then we can get the gradient formula by putting $q = p$.

Exercise B (Answers Page 132)

a Copy the table below and complete the 3rd and 4th columns.

Equation	Secant Gradient	Tangent Gradient	Put $p = x$
$y = x$	1	1	1
$y = x^2$	$p + q$	$2p$	$2x$
$y = x^3$	$p^2 + pq + q^2$		
$y = x^4$	$p^3 + p^2q + pq^2 + q^3$		
$y =$			

b Also complete the row below $y = x^4$.

c So what will be the tangent gradient for: $y = x^3$ when $x = 4$?

d And what will be the tangent gradient for: $y = x^4$ when $x = 2$?

e What is the pattern that connects the first and last columns in the table?

f What will be the tangent gradient for $y = x^7$?

g And what will be the tangent gradient for $y = x^{20}$?

h If $y = x^n$ what would the tangent gradient be?

[2] GRADIENT OF $y = ax^n$

If we have curves where the coefficient is not 1, such as $y = 2x^2$, $y = 7x^5$, we can easily get the tangent gradient too.

For example, if we have $y = 5x^3$. What is the gradient formula?

Comparing $y = 5x^3$ with $y = x^3$ we see that whatever value is chosen for x in $y = x^3$ the y-value will be 5 times bigger in $y = 5x^3$.

This is like stretching the curve to make it 5 times higher.

And the effect will be that all the gradients will be 5 times bigger too.

In other words, the gradient for any value of x in $y = 5x^3$ will 5 times the gradient in $y = x^3$.

And since the gradient formula for $y = x^3$ is $\frac{dy}{dx} = 3x^2$,

the gradient formula for $y = 5x^3$ is $\frac{dy}{dx} = 15x^2$.

And in general we can say that if $y = ax^3$ then $\frac{dy}{dx} = 3ax^2$.

More generally, if $y = x^n$ then $\frac{dy}{dx} = nx^{n-1}$,

therefore if $y = ax^n$ then $\frac{dy}{dx} = anx^{n-1}$.

EXAMPLE 1

All this simply means that to differentiate say $y = 5x^3$ **we multiply the power by the coefficient** ($3 \times 5 = 15$), **and reduce the power by 1** ($3 - 1 = 2$).

So $\frac{dy}{dx} = 15x^2$.

EXAMPLE 2

Similarly for $\boldsymbol{y = 6x^5}$ we have $\frac{dy}{dx} = \boldsymbol{30x^4}$ (where $30 = 5 \times 6$ and $4 = 5 - 1$).

In general we can say that: | if $\boldsymbol{y = ax^n}$ then $\frac{dy}{dx} = \boldsymbol{anx^{n-1}}$. |

EXAMPLE 3

For $\boldsymbol{y = 3x}$ this formula gives $\boldsymbol{3x^0}$ and since $x^0 = 1$ * we get just 3.
This makes sense because we know that the gradient of the line given by $y = 3x$ is 3.

* Since $x \div x = 1$ and also $x \div x = x^{1-1} = x^0$, it follows that $x^0 = 1$.

EXAMPLE 4

For **y = 5** we can think of 5 as $5x^0$. Then the formula gives $0 \times x^{-1}$, which equals 0. This also makes sense because $y = 5$ is a horizontal line whose gradient is zero.

Exercise C (Answers Page 132) Differentiate:

a $5x^4$	**b** $3x^3$	**c** $8x^7$	**d** x^6	**e** $2x^5$	**f** x
g $\frac{1}{2}x^4$	**h** $3x^2$	**i** $9x$	**j** $15x^4$	**k** $3x^{12}$	**l** $2x^0$

m On Page 42 we found that if $y = x^3$ then $\frac{dy}{dx} = 3x^2$.

Adapt this to show that if $y = ax^3$ then $\frac{dy}{dx} = 3ax^2$.

[3] DIFFERENTIATING SEVERAL TERMS

In Chapter 3 we saw that:

> For the quadratic $y = ax^2 + bx + c$
>
> $$\frac{dy}{dx} = 2ax + b$$

But $2ax + b$ is exactly the result we get when we apply the formula $\frac{dy}{dx} = anx^{n-1}$ to the individual terms in $y = ax^2 + bx + c$ (you can think of this as $y = ax^2 + bx^1 + cx^0$ (since $x^0 = 1$, see Chapter 7)).

This indicates that we can differentiate expressions like $5x^3 + 4x^2 - 7x + 3$ **term by term**.

That is, we differentiate each of the four terms to get $15x^2 + 8x - 7$.

[This is further explained on Page 116]

EXAMPLE 5

Differentiate $y = 7x^5 + x^4 + 9x - 8$.

- We get $\frac{dy}{dx} = 35x^4 + 4x^3 + 9$.

EXAMPLE 6

Differentiate $x^5 - 12$.

- We get $5x^4$.

EXAMPLE 7

Find $\frac{d}{dx}(2x^3 + x^2 - x)$.

- We get $6x^2 + 2x - 1$.

EXAMPLE 8

Find the gradient of the curve given by $y = x - x^3$ at the point where $x = -2$

- We get $\frac{dy}{dx} = 1 - 3x^2$.

And putting $x = -2$, $\frac{dy}{dx} = 1 - 12 = -11$.

Exercise D (Answers Page 133)

a Differentiate $3x^6 + 5x^4$

b Differentiate $77 - x^3$

c $y = 3x^4 + 4x^2 - 6x + 3$ find $\frac{dy}{dx}$

d $y = 5x - 2x^3$ find $\frac{dy}{dx}$

e Find $\frac{d}{dx}(x^5 - x^3 - x)$

f Find $\frac{d}{dx}(\frac{1}{2}x^2 - \frac{1}{2}x^3)$

g Differentiate $13x^3 + \frac{1}{2}x^4$

h Differentiate $8 + 9x^2 - x + 3x^3$

i Find $\frac{dy}{dx}$ when $y = 3$

j Find $\frac{dy}{dx}$ when $y = 50x - 50x^2$

k Use the secant-tangent method to find $\frac{dy}{dx}$ for $y = x^2 + 5x$

SECOND DERIVATIVE

$\frac{dy}{dx}$ is called the **first derivative** of y.

This is because we sometimes we get involved with the **second derivative**, which means differentiating $\frac{dy}{dx}$. So, the second derivative means $\frac{d}{dx}\left(\frac{dy}{dx}\right)$ and this is usually abbreviated to $\frac{d^2y}{dx^2}$.

EXAMPLE 9

For example, given $y = 5x^4 + 6x^2 + 7$

- We get $\frac{dy}{dx} = 20x^3 + 12x$,

  and differentiating $20x^3 + 12x$ we get $\frac{d^2y}{dx^2} = 60x^2 + 12$.

EXAMPLE 10

Given $y = 15 - 3x + 5x^2 - 7x^3$ find $\frac{d^2y}{dx^2}$.

- We find $\frac{dy}{dx}$ first: $\frac{dy}{dx} = -3 + 10x - 21x^2$.

  Therefore $\frac{d^2y}{dx^2} = 10 - 42x$.

Exercise E (Answers Page 133) Find $\frac{d^2y}{dx^2}$:

a $8x^5 + 5x^4$ **b** $77 - x + x^2 - x^3$ **c** $x^3 + x - \frac{1}{2}x^2$

d $3x^6$ **e** $1 - x^2$ **f** $5x$

One of the uses of $\frac{d^2y}{dx^2}$ is in studying acceleration, which we come to in Chapter 13. It can also tell us whether a gradient is increasing or decreasing.

Key Points of Chapter 5

* To differentiate $\boldsymbol{y = ax^n}$ we get $\boldsymbol{\frac{dy}{dx} = anx^{n-1}}$.

That is to say, given $\boldsymbol{y = ax^n}$ we multiply the power by the coefficient and reduce the power by 1.

So if $y = 6x^4$, for example, then $\frac{dy}{dx} = 24x^3$.

* We can differentiate **term by term**. So to differentiate $\boldsymbol{7x^4 + 5x^3}$ we would get $\boldsymbol{28x^3 + 15x^2}$.

* We can differentiate twice, to get the 'second derivative' denoted by $\frac{d^2y}{dx^2}$.

AHEAD

We go on next to deal with curves where x is raised to a negative or fractional value.

CHAPTER 6

Negative Powers

In this Chapter:

*** We see how to find gradients when the power is negative**

We have studied many variations of the equation $y = ax^n$ but so far n has always been a positive whole number. Here we what happens when n is negative.

THE EQUATION $y = \frac{1}{x}$

Now let's take a curve that involves a negative power: $y = \frac{1}{x}$.

We know that if we divide an expression such as $\frac{x^2}{x}$ we subtract the powers on x.

So $\frac{x^2}{x} = x^{2-1} = x$.

Similarly $\frac{x}{x^2} = x^{1-2} = x^{-1}$.

But $\frac{x}{x^2} = \frac{1}{x}$ if we cancel x.

Therefore $\frac{1}{x} = x^{-1}$.

Similarly $\frac{1}{x^2} = x^{-2}$. And so on.

In other words $y = \frac{1}{x}$ and $y = x^{-1}$ are identical.

THE CURVE $y = \frac{1}{x}$

Let us plot the graph of this equation.

Exercise A (Answers Page 133)

a Draw the graph of $y = \frac{1}{x}$ as follows.

With the origin near the centre of a sheet of graph paper and with 2 cm equal to 1 unit on both axes, draw the axes, and plot some points by completing the table below.

x =	-4	-2	-1	-½	-¼	¼	½	1	2	4
y =	-¼					4				

Join the points to get two smooth curves. Label it: $y = \frac{1}{x}$.

By carefully drawing tangents to your curve find the gradient of $y = \frac{1}{x}$ where:

b x = 2 **c** x = ½ **d** x = -2 **e** x = -½

f Use the secant-tangent method to find a formula for the gradient of this curve.

Gradient Formula Using $\frac{dy}{dx} = anx^{n-1}$

Differentiating $y = x^{-1}$ we get $\frac{dy}{dx} = -1x^{-2} = -\frac{1}{x^2}$.

> For $y = \frac{1}{x}$,
>
> $\frac{dy}{dx} = -\frac{1}{x^2}$.

Exercise B (Answers Page 134) Use this formula to find the gradient of $y = \frac{1}{x}$ at:

a x = 2 **b** x = ½ **c** x = -2 **d** x = -½

e The answers you get should be equal (or approximately equal) to the gradients you measured in the previous exercise. Check this is true.

ANOTHER CURVE: $y = \frac{1}{x^2}$

We can draw the graph of this equation.

In the table below nine *x*-values are chosen.

x =	-4	-2	-1	$-\frac{1}{2}$	0	$\frac{1}{2}$	1	2	4
y =	$\frac{1}{16}$				∞				

Infinity

For *x* = 0 we find $y = \frac{1}{0}$. This cannot be given a value in mathematics. But to get some idea what it means we can take a number that is very small, so that it is close to zero.

If you do that (say put *x* = 0.001) you will get a very large result for *y*.

And as *x* gets even smaller, *y* gets even bigger, so that we suspect that at *x* = 0 *y* will be infinitely big. This is shown by the symbol for infinity in the table: ∞.

You can't plot it on the graph, but it does show that the curve approaches the *y*-axis for small *x*-values as you will see.

Exercise C (Answers Page 134)

a Complete the bottom row in the table above. Check your answers.

b Take a sheet of graph paper and with the sheet horizontal choose a point at the bottom that is approximately in the middle.

Draw a scale of 2 cm = 1 unit on both axes.
x runs from -4 to 4,
and *y* runs from 0 to 4.

c Plot the 8 points indicated in the table above.

d Draw the curve. It will have two sections: one on each side of the *y*-axis. Label the graph: $y = \frac{1}{x^2}$.

e By drawing tangents to the curve at $x = 1$ and $x = 2$ estimate the gradient of the curve at those two points.

f What would you say is the gradient at $x = -2$?

g Use the formula $\frac{dy}{dx} = anx^{n-1}$ to get the gradient formula for $y = \frac{1}{x^2}$.

h Use the secant-tangent method to find the gradient formula for $y = \frac{1}{x^2}$. Begin by calling the two points $P(\frac{1}{p}, p^2)$ and $Q(\frac{1}{q}, q^2)$.

i Confirm that the gradient formula gives the same gradient values at $x = 1$ and $x = 2$ as you estimated in **e** above.

OTHER CURVES WITH NEGATIVE *n*

EXAMPLE 1

Find the gradient of the curve given by $y = \frac{2}{x}$ at the point where x = 3.

- We have $y = \frac{2}{x} = 2x^{-1}$.

  And since for $y = ax^n$, $\frac{dy}{dx} = anx^{n-1}$

  therefore $\frac{dy}{dx} = -2x^{-2} = \frac{-2}{x^2}$.

  And if $x = 3$, $\frac{dy}{dx} = -\frac{2}{9}$.

EXAMPLE 2

Find $\frac{dy}{dx}$ at $x = 2$ for $y = \frac{1}{3x^2}$.

- We have $y = \frac{1}{3x^2} = \frac{1}{3}x^{-2}$.

  Therefore $\frac{dy}{dx} = \frac{-2}{3}x^{-3} = \frac{-2}{3x^3}$.

  And if $x = 2$, $\frac{dy}{dx} = -\frac{1}{12}$.

<u>**Exercise D**</u> (Answers Page 134) Use $\frac{dy}{dx} = anx^{n-1}$ to differentiate:

a $y = \frac{3}{x^2}$ **b** $y = \frac{1}{3x}$ **c** $y = \frac{7}{x^3}$

d Find the gradient of $y = \frac{2}{x^2}$ at the point where $x = -1$

e Find the gradient of $y = \frac{3}{x^3}$ at the point (-1, -3)

Key Points of Chapter 7

* We can draw curves of the type $y = ax^n$, where n is negative.

* We can get exact gradients of these curves by using the secant-tangent method or the formula $\frac{dy}{dx} = anx^{n-1}$.

AHEAD

Next we see a really easy way to find gradients of curves using a simple ratio.

CHAPTER 7

Fractional Powers

In this Chapter:

*** We see how to find gradients when the power is a fraction.**

Here we see what happens when n is a fraction, like $\frac{1}{2}$.

THE EQUATION $y^2 = x$

Suppose we have the equation $y^2 = x$.

To understand better what we are dealing with we can draw the graph of $y^2 = x$.

The graph consists of all the points that satisfy the equation $y^2 = x$.

For example the point (9, -3) will be on the curve because it is true that $y^2 = x$ when $x = 9$ and $y = -3$. The coordinates satisfy the equation.

Since x equals y^2 it makes sense to choose the y-values and square them to find the corresponding x-values.
So we could choose the y-values shown below:

$y =$	-3	-2	-1	0	1	2	3
$x =$	9		1	0	1		

Exercise A (Answers Page 134, 5)

a Fill in the three missing numbers in the table above.

b Plot these 7 points on a graph. With the graph sheet horizontal, put the origin near the middle on the left. You can use a scale of 2 cm = 1 unit on both axes.

You could also plot a few more points to get a more accurate graph. Since $x = y^2$ you could choose, for example, $y = 2.5$, and square that to get the x-value. So you would plot (6.25, 2.5). And so on for some other y-values.

c Fill in the two missing numbers below and plot 6 more points:

$y =$	-2.5	-1.5	-0.5	0.5	1.5	2.5
$x =$	6.25		0.25	0.25		6.25

d Join the points to get a smooth curve. Label it: $y^2 = x$.

e Draw the two tangents at $x = 1$, and measure their gradients.

f Draw the two tangents at $x = 4$, and measure their gradients.

g Use the secant-tangent method to find a gradient formula for $y^2 = x$. Begin by calling the two points $P(p^2, p)$ and $Q(q^2, q)$.

Keep your graph for later.

Making *y* the subject

To make y the subject in $y^2 = x$ we take the square root of both sides.

We get $y = \pm\sqrt{x}$.

The ± symbol here means there are two answers, one plus and one minus.

If we had $y^2 = 25$ we would get $y = +5$ or $y = -5$ because both +5 and -5 give 25 when squared.
So we would say, if $y^2 = 25$ then $y = \pm 5$.

In the same way $y^2 = x$ gives $y = \pm\sqrt{x}$.

A power of $\frac{1}{2}$ means a square root

We note that $\sqrt{x} \times \sqrt{x} = x$.

But $x^{\frac{1}{2}} \times x^{\frac{1}{2}} = x^1 = x$.

It follows that a square root is equivalent to a power of $\frac{1}{2}$.

So we find that if $y^2 = x$ then $y = \pm\sqrt{x}$ becomes $y = \pm x^{\frac{1}{2}}$.

We find the power on x is a fraction: $\frac{1}{2}$.

[The 'Laws of Indices' can be found on Page 117 for reference]

Differentiating $y^2 = x$

We obtained the gradient formula in Exercise A, Question **g**.

But we can also get the gradient formula for $y^2 = x$ by using the formula from Chapter 5:

For $y = ax^n$

$$\frac{dy}{dx} = anx^{n-1}.$$

This formula works when n is a fraction or negative as well as for positive whole numbers.

Differentiating $y = \pm x^{\frac{1}{2}}$ we get $\frac{dy}{dx} = \pm\frac{1}{2}x^{-\frac{1}{2}}$, which is the same as $\pm\frac{1}{2\sqrt{x}}$.

For $y^2 = x$

$$\frac{dy}{dx} = \pm\frac{1}{2\sqrt{x}}.$$

Exercise B (Answers Page 136)

a Put $x = 1$ in the gradient formula just found and calculate the gradient at the two points where $x = 1$.
(You will get two answers because the square root of 1 is +1 or -1.)

b Compare these numbers with the gradients you measured earlier at $x = 1$ (in Exercise A, question **e**).

c Put $x = 4$ in the gradient formula and calculate the gradients of two more points.

d Compare these numbers with the gradients you measured earlier at $x = 4$ (in Exercise A, question **f**).

Keep your graph for later.

ANOTHER CURVE: $y^2 = x^3$

Given the equation $y^2 = x^3$, by taking the square root of both sides we get $y = \pm x^{\frac{3}{2}}$.
So here the fractional power is $\frac{3}{2}$.

You can plot a few points on the curve to get an idea of its shape.

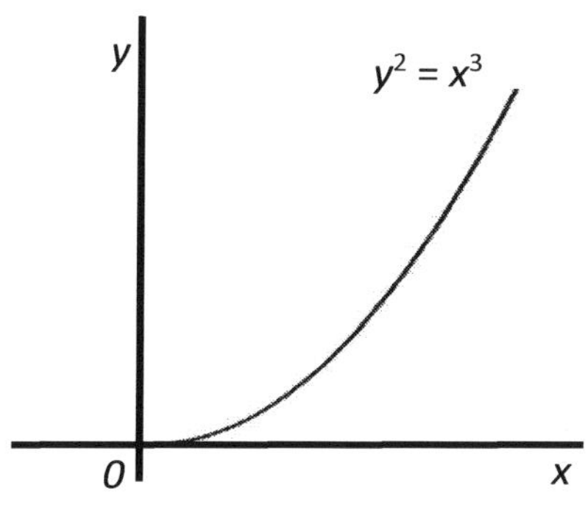

The table below shows seven points that lie on the curve.

$y =$	-27	-8	-1	0	1	8	27
$x =$	9	4	1	0	1	4	9
$y^2 =$	729						
$x^3 =$	729						

Exercise C (Answers Page 136)

a To confirm that the points are correct fill in the bottom two rows of the table above. This should prove that $y^2 = x^3$ for those points. The first column has been completed for you.

b Take a sheet of graph paper and with the sheet vertical choose a point on the left side and half way up.
Draw a scale of 2 cm = 2 units on the x-axis and 2 cm = 5 units of the y-axis.
x runs from 0 to 9,
and y runs from -27 to 27 (you can mark every 5 on the vertical axis).

c Plot the 7 points indicated in the table above.

d Plot the point (6.25, 15.625) as well.

e Draw the curve. It will have a point at the origin and the x-axis will be a line of symmetry. Label it: $y^2 = x^3$.

The Gradient for $y^2 = x^3$

Now applying our formula, $\frac{dy}{dx} = anx^{n-1}$:

$$\text{If } y = \pm x^{\frac{3}{2}}$$
$$\text{then } \frac{dy}{dx} = \pm \tfrac{3}{2} x^{\frac{1}{2}}$$

Exercise D (Answers Page 136)

a Use this formula to calculate the two gradients of the curve where $x = 4$.

b Measure the gradients on your graph at $x = 4$.

c Use the formula to calculate the gradients of the curve where $x = 1$.

d Measure the gradients on your graph at $x = 1$.

e Measure the gradients on your graph at $x = 2$.

f Use the secant-tangent method to derive the result: $\frac{dy}{dx} = \pm\frac{3}{2}x^{\frac{1}{2}}$. Begin by calling the two points $P(p^2, p^3)$ and $Q(q^2, q^3)$.

<u>**Exercise E**</u> (Answers Page 137) Use $\frac{dy}{dx} = anx^{n-1}$ to differentiate:

a $y = x^{\frac{5}{2}}$ **b** $y = 2x^{\frac{5}{2}}$ **c** $y = 10x^{\frac{2}{5}}$ **d** $y^3 = x^2$

Find the gradient of:

e $y^3 = x^2$ at the point (8, 4)

f $y^5 = x^2$ at the point (1, 1)

g $y^2 = x^5$ at the point where $x = 4$

Key Points of Chapter 6

* We can draw curves of the type $y = ax^n$, where n is a fraction.

* We can get exact gradients of these curves by using $\frac{dy}{dx} = anx^{n-1}$.

This formula applies where n is fractional or negative (or both).

AHEAD

Next we look at curves for which n is negative in $y = ax^n$.

CHAPTER 8

Ratio of Gradients

In this Chapter:

*** We see that for the curves $y = ax^n$ there is a ratio of gradients that is equal to the power, *n*.**

*** And how we can use that ratio to get gradients.**

THE CURVE $y = x^2$

We saw in Chapter 2 that the gradient of a secant for $y = x^2$ is $p + q$: just the **sum of the *x*-coordinates** of the points where the secant and curve meet.

So in the diagram opposite

the secant gradient is -2 + 5 = 3.

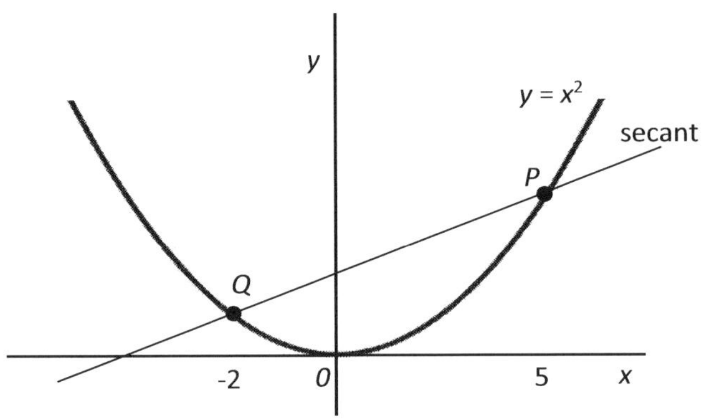

But if *Q* moves along the curve towards *P* it will reach the origin at some point, and the sum of the *x*-coordinates will then be 0 + 5 = 5.

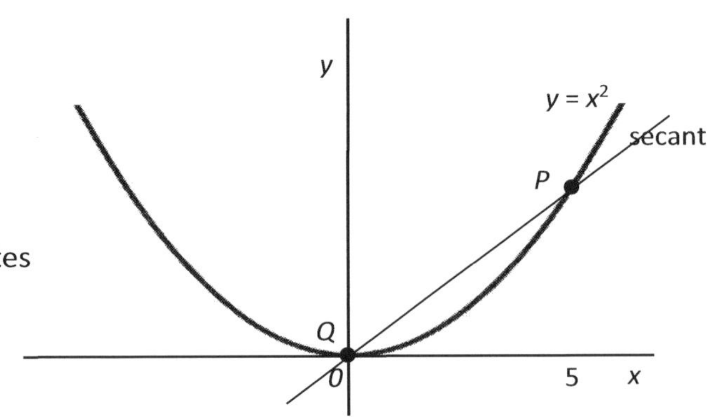

And in fact the gradient of the origin secant will always be just the x-coordinate of P.

We can also put $q = 0$ in $p + q$ we get p, which is the x-coordinate of P.

When Q arrives at P, and the secant becomes a tangent, the two x-coordinates will be the same, so $p + q = 2p$ and we get a tangent gradient of $2p$ or $2x$.

This was discussed at the beginning of Chapter 3, and these results are summarised in the diagram below.

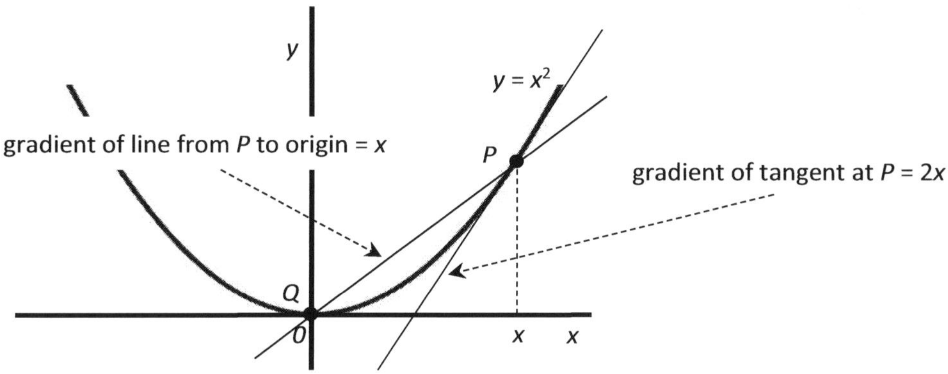

This means that the tangent gradient is always twice the origin gradient, for this curve.

<u>**Exercise A**</u> (Answers Page 137)

a Draw an accurate graph of the curve $y = x^2$.
(You may have the graph you used in Chapter 3, Exercise A.)

b Use your graph to confirm that, for the point (3, 9) on the curve, the tangent gradient is twice the origin gradient.

c Check also that for the point (-2, 4) that the tangent gradient is twice the origin gradient.

d Locate the point on the curve where $x = 1.5$ and show that the tangent gradient is twice the origin gradient.

e Can you find a point where the tangent gradient is not twice the origin gradient?

THE CURVE $y = x^3$

Similarly for $y = x^3$ we had: **secant gradient** = $\boldsymbol{p^2 + pq + q^2}$ (shown on Page 42):

Putting $q = 0$ we get the origin gradient = p^2.

And putting $q = p$ we get the tangent gradient = $3p^2$.

We note that the tangent gradient is 3 times the origin gradient this time.

THE CURVES $y = ax^n$

Note that these examples, $y = x^2$ and $y = x^3$, suggest the gradient tangent is n times the origin gradient.

And it is in fact true that for $\boldsymbol{y = ax^n}$ the tangent gradient is n times the origin gradient.

This is proved on Page 117,8.

To put it another way: **the ratio of gradients is $\boldsymbol{n}$**.

Later you will see another, similar formula: that a **ratio of areas** equals n, which is very useful for finding areas under curves.

EXAMPLE 1

What is the ratio of gradients for $y = 5x^3$?

- Comparing $5x^3$ with ax^n we see that $n = 3$. So the ratio of gradients is 3.

(The tangent gradient is 3 times to origin gradient.)

Notice that the value of a makes no difference: whatever value a has, the ratio of gradients will be 3.

EXAMPLE 2

What is the ratio of gradients for $y^2 = 5x^3$?

We would need to take the square root of both sides in order to make y the subject, which would mean the power on x is $\frac{3}{2}$.

Exercise B (Answers Page 137)

What do you expect the ratio of gradients will be for:

a $y = 3x^2$ **b** $y = 3x^5$ **c** $y = \frac{4}{x}$ **d** $y^2 = x^5$

e $x^2 = 7y^3$ **f** $y^2 = 2x$ **g** $y = \frac{2}{3x^2}$

RATIO OF GRADIENTS = *n*

As mentioned before, for $y = ax^n$ **the ratio of gradients is *n***.

If P has coordinates (x, y) the origin gradient will be $\frac{y}{x}$. The tangent gradient is $\frac{dy}{dx}$.

So we can write: $\dfrac{dy/dx}{y/x} = n$.

EXAMPLE 3

By finding $\frac{dy}{dx}$ and $\frac{y}{x}$ show that for $y = 5x^3$ the ratio of gradients at the point (2, 40) equals 3.

- First note that for $y = 5x^3$ the value of $n = 3$

  We can confirm that (2, 40) is a point on the curve, because $40 = 5\times2^3$.

  The tangent gradient is given by $15x^2$ and if $x = 2$, then $\frac{dy}{dx}$ = **60**.

  And $\frac{y}{x} = \frac{40}{2}$ = **20**.

  And we note that the ratio of 60 to 20 is **3**.

EXAMPLE 4

By finding $\frac{dy}{dx}$ and $\frac{y}{x}$ show that for $y = 5x^3$ the ratio of gradients equals 3 at **any point** on the curve.

- For **any point** on the curve, we need to work with $\mathbf{5x^3}$.

  We know that $\frac{dy}{dx} = 15x^2$.

  and $\frac{y}{x} = \frac{5x^3}{x} = \mathbf{5x^2}$.

  So $\frac{dy/dx}{y/x} = \frac{15x^2}{5x^2} = \mathbf{3}$.

A NEGATIVE POWER

<u>**Exercise C**</u> (Answers Page 137)

You will need the graphs you drew earlier here.

a Draw an accurate graph of the curve $y = \frac{1}{x}$ (or use one from earlier).

b Use your graph to confirm that for the point (1, 1) on the curve the ratio of gradients is -1.

c Confirm for the points $(2, \frac{1}{2})$, $(-2, -\frac{1}{2})$ and $(3, \frac{1}{3})$ that the ratio gradients is -1.

d Using your graph for $y^2 = x$ confirm that the ratio of gradients is $\frac{1}{2}$ for the point (9, 3).

e Using your graph for $y^2 = x^3$ confirm that the ratio of gradients is $\frac{3}{2}$ for the point (4, -8).

f Using your graph of $y = \frac{1}{x^2}$, confirm for the points (1, 1) and (2, ¼), that the ratio of gradients is -2.

g Using these graphs (and the one for $y = x^2$ earlier) you can continue to draw more pairs of gradients and construct some unusual and pleasing patterns.

FINDING GRADIENTS USING THE RATIO

We can use our ratio to get gradients of curves. Since the ratio of gradients is *n*:

$$\frac{dy}{dx} = n \times \frac{y}{x}.$$

This can be easier to use than $\frac{dy}{dx} = anx^{n-1}$.

$\frac{dy}{dx} = anx^{n-1}$ is best used when *n* is a positive integer.

But when *n* is fractional or negative the formula $\frac{dy}{dx} = n \times \frac{y}{x}$ is better.

EXAMPLE 5

Find the gradient of the curve given by $y = \frac{3}{x^2}$ at the point (1,3).

- We see here that $n = -2$, so $\frac{dy}{dx} = -2 \times \frac{3}{1} = -6$.

EXAMPLE 6

Find the gradient of the curve given by $y^2 = 2x^3$ at the point (2, 4).

- We see here that $n = \frac{3}{2}$, so $\frac{dy}{dx} = \frac{3}{2} \times \frac{4}{2} = 3$.

EXAMPLE 7

Find the gradient of the curve given by $y^2 = 3x$ where x = 3.

- We have $n = \frac{1}{2}$, and if $x = 3$ then $y = +3$ or -3.
  So here we have two points: (3, 3) and (3, -3).

  For (3, 3) $\frac{dy}{dx} = \frac{1}{2} \times \frac{3}{3} = \frac{1}{2}$.

  For (3, -3) $\frac{dy}{dx} = \frac{1}{2} \times \frac{-3}{3} = -\frac{1}{2}$.

Exercise D (Answers Page 137) Find the gradient of the curve given by:

a $y = \frac{4}{x^2}$ at the point (2, 1)

b $y = \frac{5}{x}$ at the point (1, 5)

c $y = \frac{2}{x^3}$ at the point (3, $\frac{2}{27}$)

d $y^2 = 2x^3$ at the point (2, 4)

e $y = 3x^{-2}$ at the point (-3, $\frac{1}{3}$)

f $y^3 = x^2$ where x = -1

g $y^2 = x^3$ where y = -1

Key Points of Chapter 8

* For $y = ax^n$, the ratio of gradients (tangent gradient to origin gradient) equals n.

* This means that $\frac{dy}{dx}$ can also be found using $\frac{dy}{dx} = n \times \frac{y}{x}$, where (x, y) is a point on $y = ax^n$.

* And this is especially useful when n is fractional or negative.

AHEAD

Next we move on and see how to find an area under a parabola. This is developed further in later chapters and linked to the reverse of differentiation in Chapter 12. That is because differentiation and its reverse are closely connected: differentiation gives gradients of curves and its reverse (called integration) gives areas under them.

CHAPTER 9

Area Under a Parabola

In this Chapter:
*** How we can easily get areas under a parabola**
*** How to use symmetry to extend the method**

RATIOS OF AREAS

Take a pair of axes and from the origin draw a straight line as shown below.

Now select any point on the line and drop perpendiculars onto the two axes, as also shown.

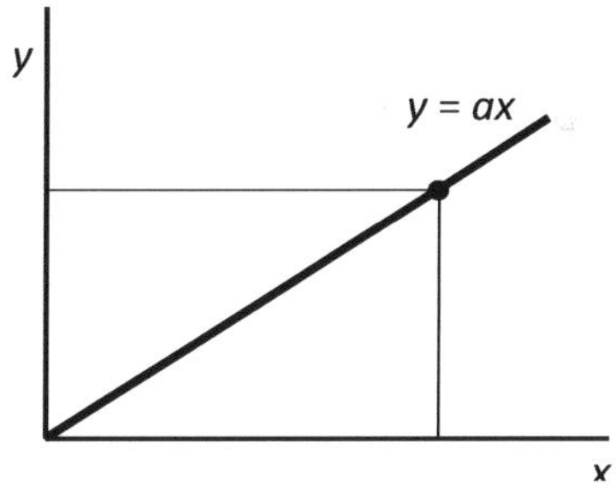

The two triangles generated will be equal in area you will agree. And this will apply for any straight line chosen: the ratio of areas will be 1:1.

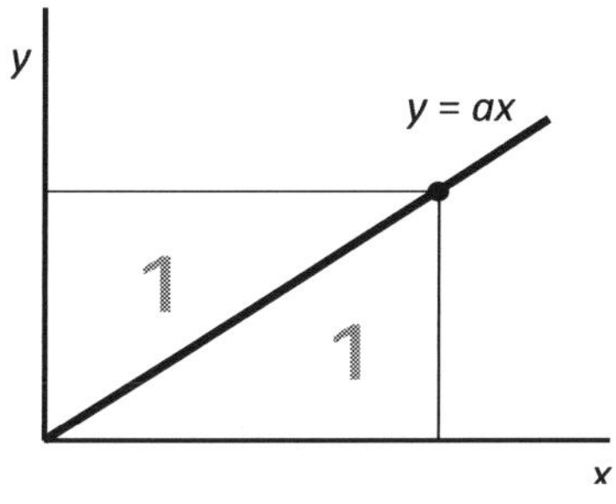

Now suppose we wish to draw a line, starting at the origin, but the requirement is that the ratio of areas similarly generated will be 2:1 rather than 1:1.

Think about it.

You will realise that the line must be a curve, not a straight line.

Also its gradient must be increasing if the area to the left of the line is to be double the area under it.

What will be the equation of the curve that has this special property?

Well, it turns out that the answer is the parabola given by $y = x^2$.

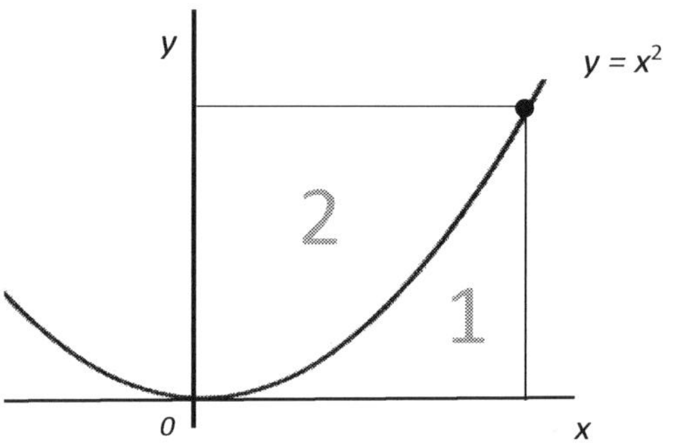

In fact any of the parabolas given by $y = ax^2$ will have this property.
[This result is proved on Pages 118-121]

This means that the area indicated by the '1' above is one third of the total area, and area '2' will be two thirds of the total rectangle.

EXAMPLE 1

Suppose we want the area under $y = x^2$ from $x = 0$ up to $x = 4$.

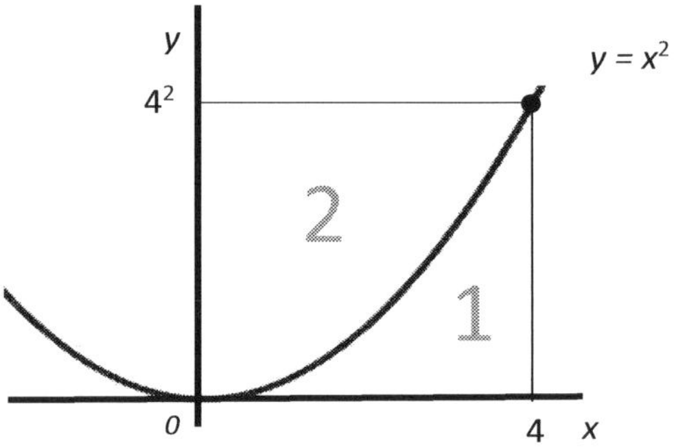

We can find the area of the rectangle and find one third of that.

The height of the rectangle is $4^2 = 16$, so the rectangle area is $4 \times 16 = 64$.

And the required area will therefore be $\frac{1}{3} \times 64 = \frac{64}{3}$.

Exercise A (Answers Page 137) Find the area under $y = x^2$ from:

a $x = 0$ to $x = 1$ **b** $x = 0$ to $x = 6$ **c** $x = 0$ to $x = ½$

d $x = 0$ to $x = 2$ **e** $x = 0$ to $x = 10$

f Find the area A:

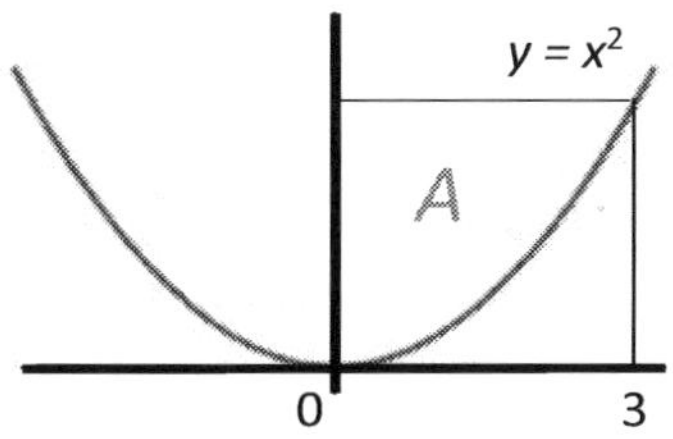

g Find the area B:

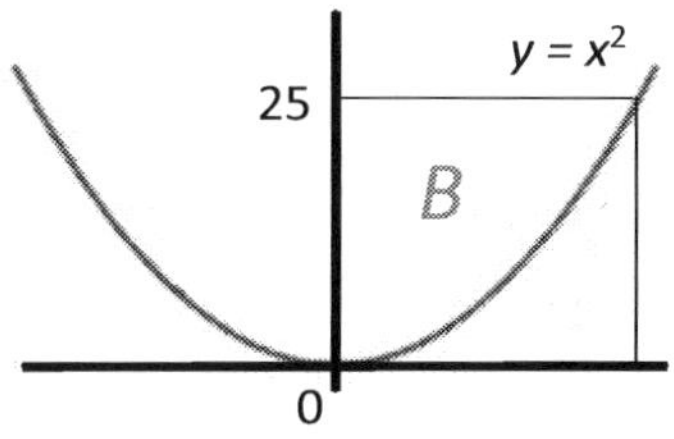

h Using your graph of $y = x^2$ from earlier verify by counting squares that the area A in **f** above is twice the area under the curve.

i Find the area enclosed by the line $y = x$ and the curve $y = x^2$.

AREA OF A STRIP

EXAMPLE 2

Suppose we want the area of a strip like the area A from $x = 2$ to $x = 3$ shown below.

How could we find it?

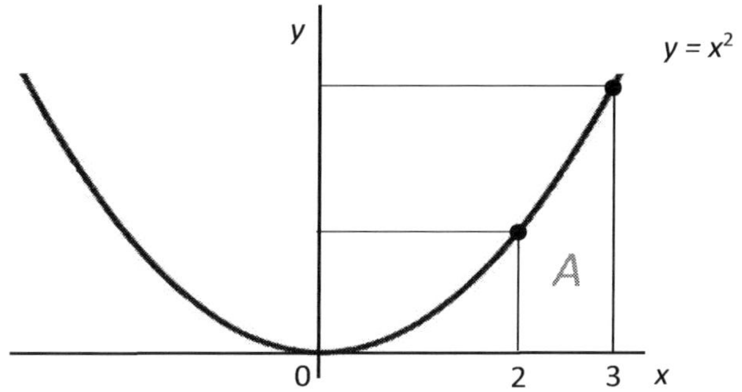

Since we can find the area from the origin up to $x = 3$ and also the area up to $x = 2$, we can find the difference of these, which will be the required area.

So we need $\frac{3^3}{3} - \frac{2^3}{3} = \frac{27}{3} - \frac{8}{3} = \frac{19}{3}$.

We get the area as $\frac{19}{3}$.

The Easiest Way

The easiest way to do this is to cube the two *x*-values, subtract them and then divide by 3.

So $A = \frac{1}{3}(3^3 - 2^3) = \frac{19}{3}$.

Exercise B (Answers Page 138)

Find the area of the strip shown:

a

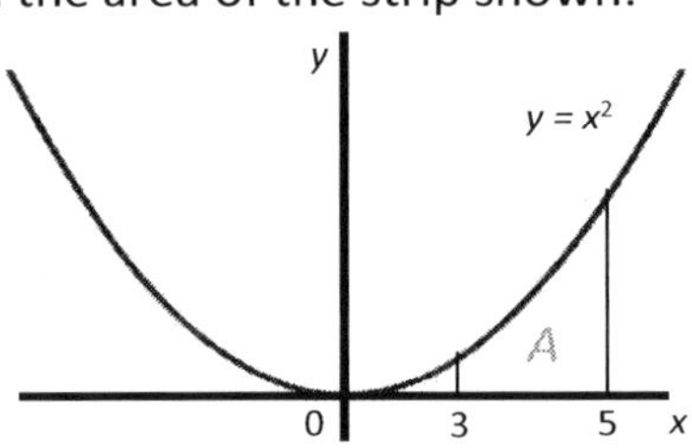

b

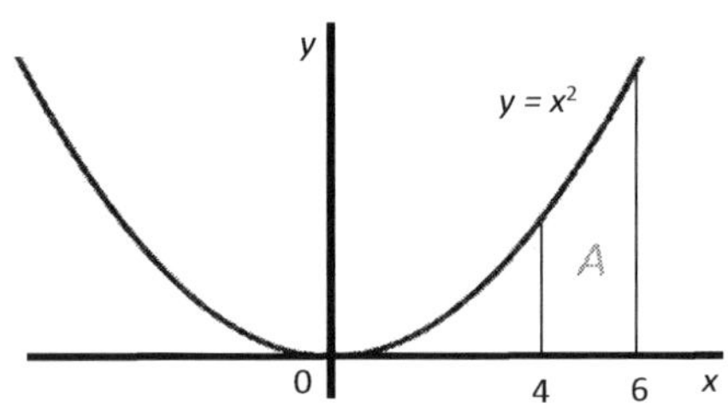

Find ... x-axis and the ...

c x ...

e Fi...

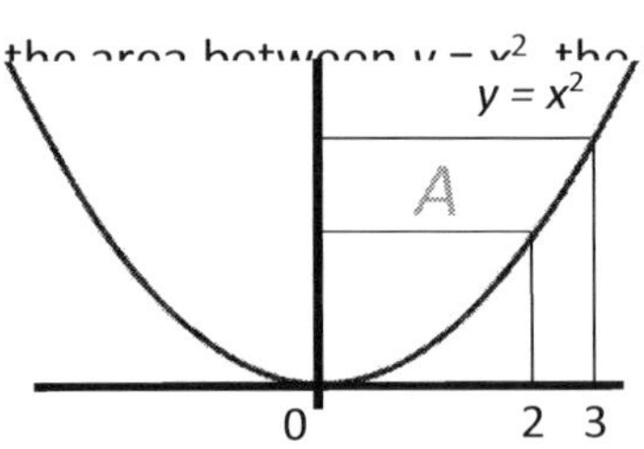

d x ...

f Fi...

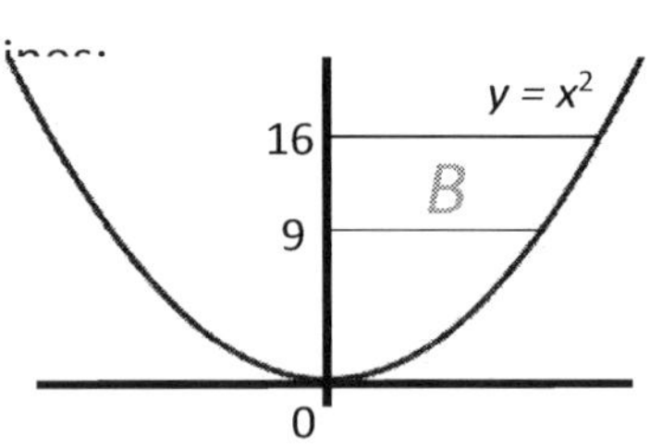

USING SYMMETRY

Because the parabola $y = x^2$ is symmetrical about the *y*-axis, and $y = -x^2$ is the reflection of $y = x^2$ in the *x*-axis, we can use symmetry to get other areas, as the next two examples show.

EXAMPLE 3

Let's look at the area under $y = x^2$ between $x = -1$ and $x = -3$, as shown.

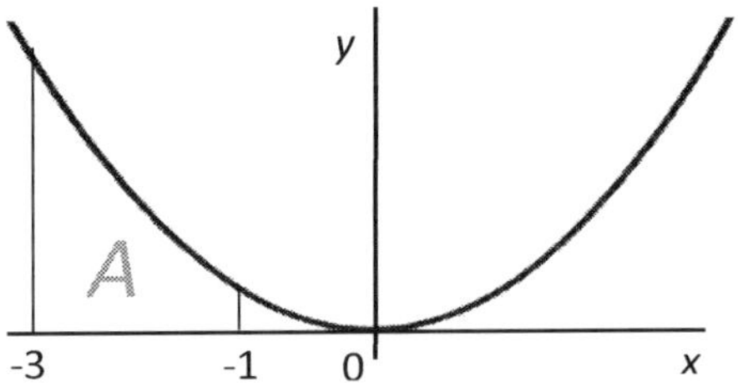

- Since the parabola is symmetrical about the y-axis the area must be the same as the area under the curve between $x = 1$ and $x = 3$ as shown.

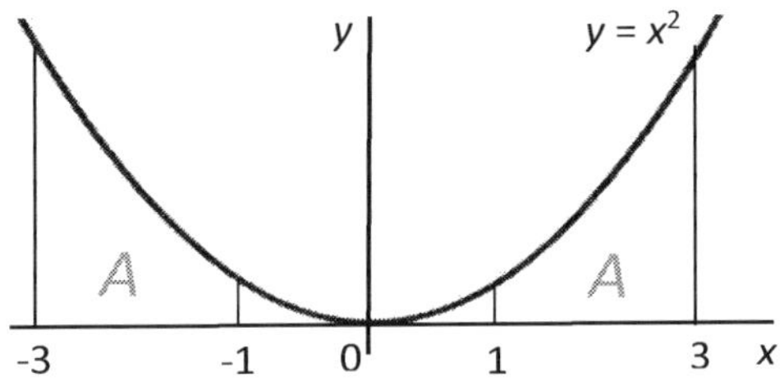

We have previously found that area to be $\frac{26}{3}$ and so that is the answer.

EXAMPLE 4

Find the area between the curve $y = -x^2$, the x-axis and the lines $x = 1$ and $x = 2$.

- Here the parabola is inverted and the required area is as shown.

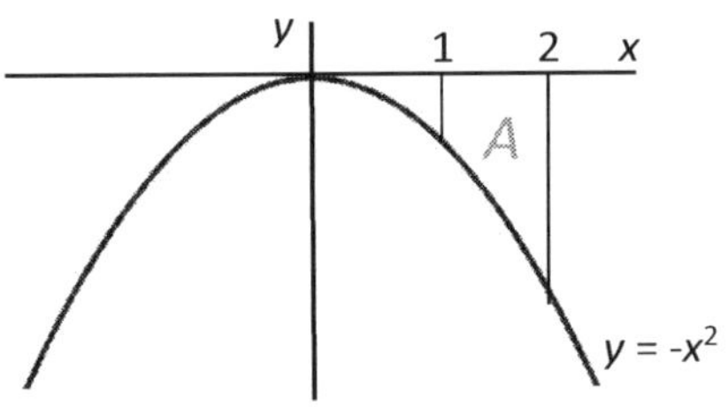

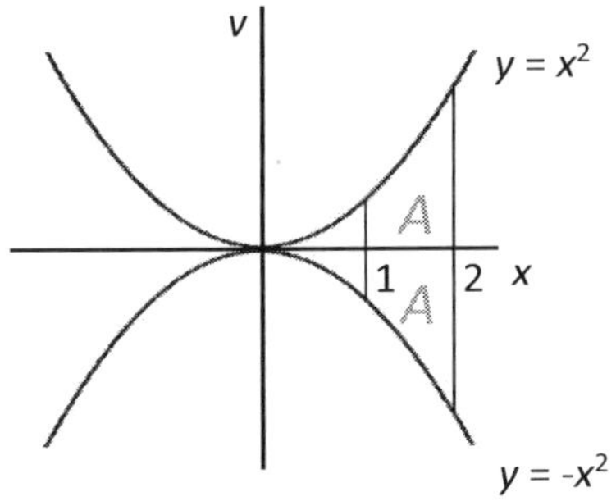

Comparing with $y = x^2$ we see we can get the answer by finding the area under $y = x^2$ between $x = 1$ and $x = 2$.

The two curves and two areas are shown together on the left.

- So we can find the area under $y = x^2$ between $x = 1$ and $x = 2$.
  $\frac{1}{3}(2^3 - 1^3) = \frac{7}{3}$.

EXAMPLE 5

Find the area under $y = x^2$ between $x = -1$and $x = 0$, and also between $x = 0$ and $x = 2$.

These areas are shown below and are $\frac{1}{3}$ and $\frac{8}{3}$.

If we add these we get $\frac{9}{3} = 3$.

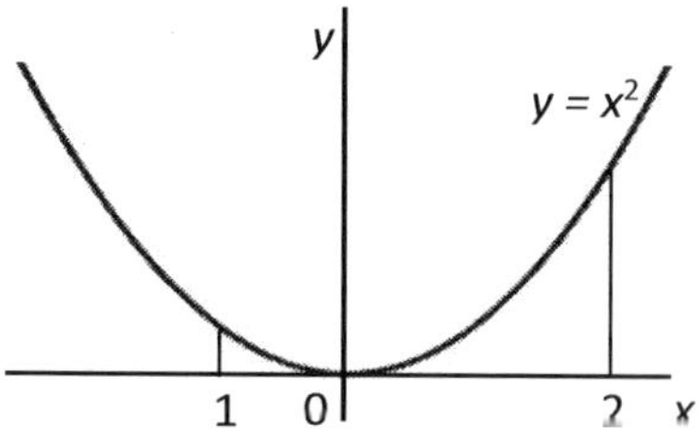

But if we calculate $\frac{1}{3}$ of the difference of the cubes, as we have been doing, we get:
$\frac{1}{3}(2^3 - (-1)^3) = 3$.

In other words we can get the full area, from $x = 2$ to $x = -1$, in one go by finding $\frac{1}{3}$ of the difference of the cubes (greater cube minus lesser cube).

Exercise C (Answers Page 138) Find the area under $y = x^2$:

a between $x = -3$ and $x = -2$, **b** between $x = -4$ and $x = -1$

c Find the area shown opposite.

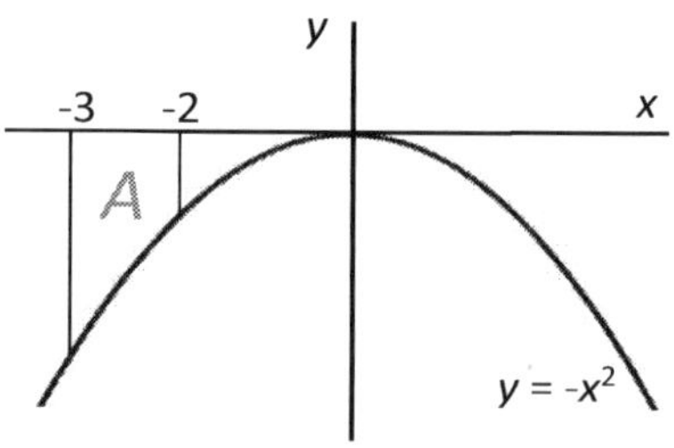

d Draw a single sketch showing the curves $y = x^2$ and $y = -x^2$. Find the total area enclosed between these curves and the lines $x = 1$ and $x = 3$.

$y = ax^2$

If the curve was $y = ax^2$ the a has the effect of stretching the curve vertically from the x-axis.

So if $a = 2$, for example, then we have $y = 2x^2$, and the curve is stretched vertically by a factor of 2. So the areas will get doubled (but the ratio of areas will be unchanged).

This is illustrated in the diagrams opposite.

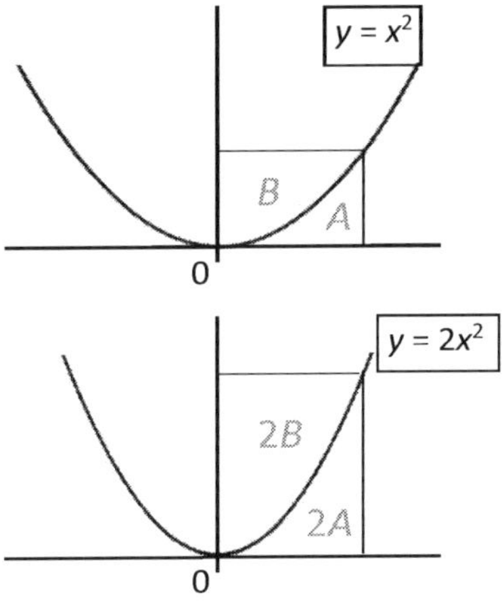

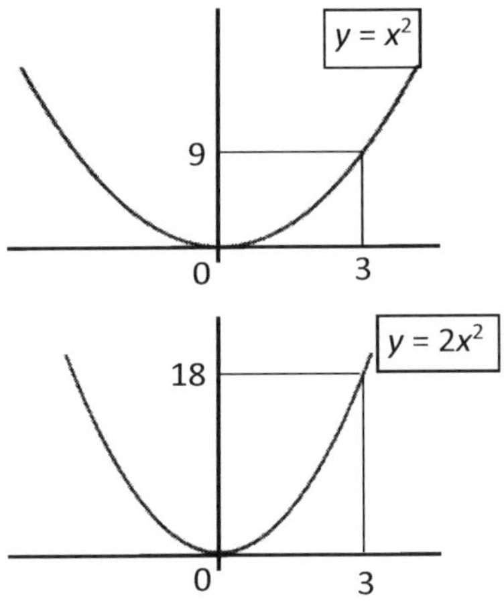

To illustrate further, if the x-coordinate was 3 the graphs would look as shown on the left here.

That is to say we can find the difference of the cubes of the x-limits, divide by 3 **and multiply by a**.

If a is negative we simply get a negative answer and we can drop the minus as area is always positive (see Example below).

To get the area under $y = ax^2$ find the area under $y = x^2$ and multiply by a.

EXAMPLE 6

Find the area under $y = 4x^2$ between $x = 1$ and $x = 2$.

- We find the difference of the cubes of 1 and 2, divide by 3 **and multiply by 4**:

  $\frac{4}{3}(2^3 - 1^3) = \frac{28}{3}$.

EXAMPLE 7

Find the area under $y = ½x^2$ between $x = -2$ and $x = 0$.

- This will be the same as the area under $y = ½x^2$ between $x = 0$ and $x = 2$.

  We find the difference of the cubes of 2 and 0, divide by 3 **and multiply by ½**:

  $\frac{1}{6}(2^3 - 0^3) = \frac{4}{3}$.

Alternatively, instead of using the reflection, we can just use the limits given, $x = -2$ and $x = 0$, and find the greater cube minus lesser cube:

$\frac{1}{6}(0^3 - (-2)^3) = \frac{4}{3}$.

Note that we consistently find the cube of the greater x-value minus the cube of the smaller x-value.

EXAMPLE 8

Find the area between $y = -3x^2$, $x = 1$, $x = 4$ and the x-axis.

- We can find the area over $y = 3x^2$ between $x = 1$ and $x = 4$.

  We find the difference of the cubes of 4 and 1, divide by 3 and multiply by 3:

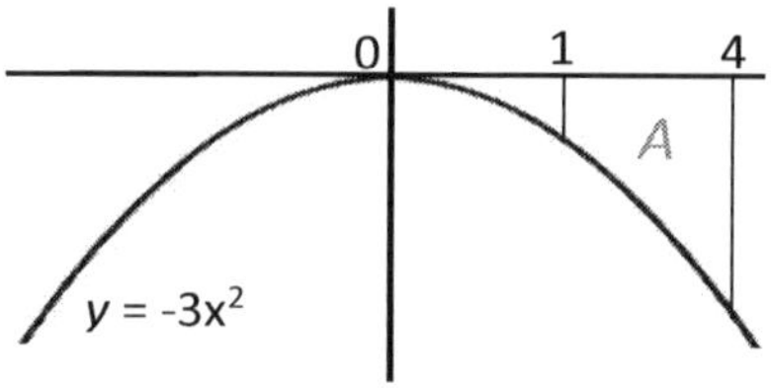

$A = \frac{3}{3}(4^3 - 1^3) = 63.$

The area is **63**.

- Alternatively if we apply our formula directly to $y = -3x^2$ we get the same result except that we multiply by -3, instead of 3:

  We get $\frac{-3}{3}(4^3 - 1^3) = -63.$

The significance of the minus result here is that it indicates an area below the x-axis.

<u>Exercise D</u> (Answers Page 138) Find the area under:

a $y = 3x^2$ between $x = 1$ and $x = 2$

b $y = 2x^2$ between $x = 2$, $x = 3$

c $y = 5x^2$ between $x = 1$ and $x = 4$

d $y = 2x^2$ between $x = -2$ and $x = 3$

e $y = 2x^2$ between $x = -2$ and $x = -3$

f $y = -x^2$ between $x = -1$ and $x = -2$

g Sketch the curves $y = x^2$ and $y = 2x^2$. Find the area enclosed between these curves and the lines $x = 1$ and $x = 2$.

h Sketch the curves $y = x^2$ and $y = -2x^2$. Find the area enclosed between these curves and the lines $x = 1$ and $x = 2$.

Key Points of Chapter 9

* For $y = ax^2$ the ratio of areas against the y and x-axes is 2:1 and we can use this fact to get the area of a section under the curve by finding one third of the full rectangular area.

* We can make use of symmetry when the x-limits are negative or if a is negative in $y = ax^2$.

* In finding the areas we consistently find the cube of the greater x-limit minus the cube of the lesser x-limit. Then multiply by a and divide by 3.

* A negative result indicates the area found is below the x-axis.

AHEAD

Next we extend to find areas under cubic and other curves.

CHAPTER 10

Ratio of Areas

In this Chapter:

* **How to get the area under a straight line, sloping or horizontal**
* **How to get the area under a cubic curve**
* **How the various results are unified**
* **How to get the area under a curve which is defined by several terms**

DIFFERENCE OF SQUARES OF *x*-LIMITS

Now let us return to the beginning of Chapter 9 where we discussed the ratio of areas for a straight line through the origin. We said the ratio of areas was 1:1.

Suppose we want the area shown, under the line $y = 2x$ between $x = 3$ and $x = 5$.

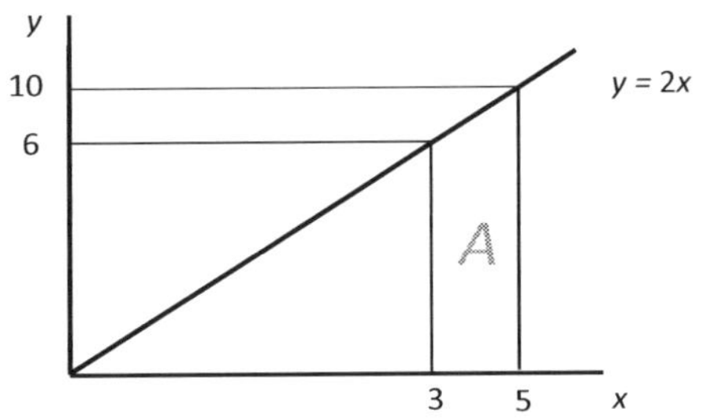

We could find the two y-coordinates, find the two rectangular areas, and then subtract them and divide by 2.

We get ½(50 – 18) = 16.

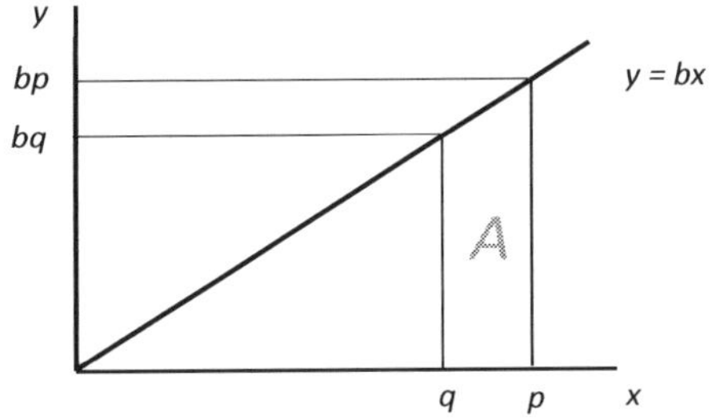

If the equation of the line is $\boldsymbol{y = bx}$, as shown, we can follow the same procedure and get:

$A = \frac{1}{2}(bp^2 - bq^2) = \frac{b}{2}(p^2 - q^2)$.

That is to say, we **subtract the squares of the *x*-limits, multiply by *b* and divide by 2**.

This is very similar to what we had for the area under $y = ax^2$, which was:

subtract the cubes of the *x*-limits, multiply by *a* and divide by 3.

To summarise:

Sloping line: $y = bx$, $A = \frac{b}{2}(p^2 - q^2)$.

Parabola: $y = ax^2$, $A = \frac{a}{3}(p^3 - q^3)$.

EXAMPLE 1

Find the area under $y = 5x$ between $x = 2$ and $x = 6$.

- We subtract the squares of 6 and 2, multiply by 5 and divide by 2:

$A = \frac{5}{2}(6^2 - 2^2) = \mathbf{80}$.

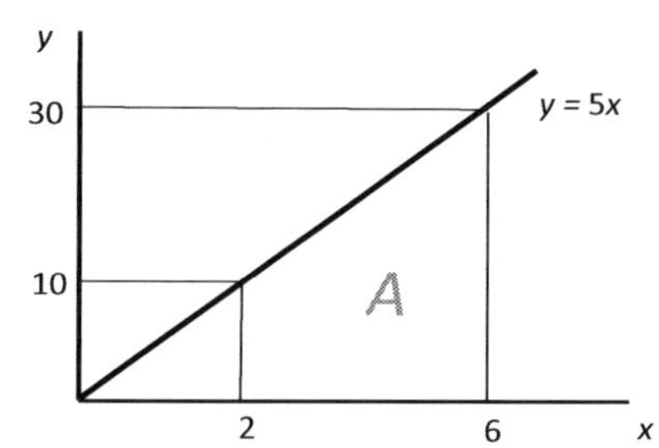

Exercise A (Answers Page 138)

a The diagram above illustrates Example 1. Confirm the *y*-values shown, and that 80 is correct as half the difference of the two rectangular areas.

Find the area under:

b $y = 3x$ between $x = 4$ and $x = 6$

c $y = 2x$ between $x = 1$ and $x = 7$

d $y = -2x$ between $x = 3$ and $x = 7$

e $y = \frac{1}{2}x$ between $x = 2$ and $x = 3$

DIFFERENCE OF *x*-LIMITS

Let's look at a horizontal line next.

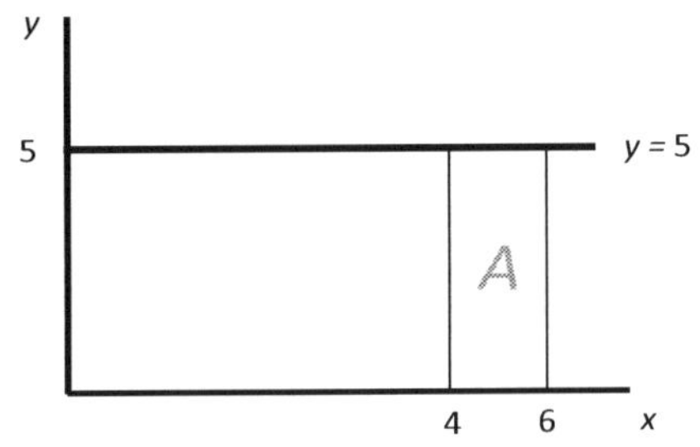

Suppose we need the area under $y = 5$ from $x = 4$ to $x = 6$, as shown.

It will be the difference of the areas of the two overlapping rectangles:

$30 - 20 =$ **10**.

Now suppose we had $\boldsymbol{y = c}$.

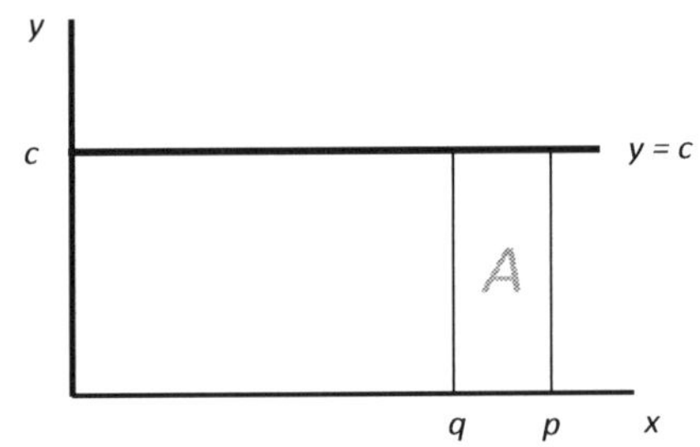

If we want the area below $y = c$ between the x-limits p and q, we can subtract the area of the rectangle up to the $x = q$ line from the rectangle up to the $x = p$ line.

We get $A = pc - qc = c(p - q)$.

So to find the area under a line $y = c$, we subtract the x-limits and multiply by c.

EXAMPLE 2

Find the area under $y = 4$ between $x = 5$ and $x = 8$.

- We find $4(8 - 5) =$ **12**.

SUMMARY

In fact if you look carefully you will see that we have a pattern:

	Equation	**Area in 1st quadrant, from $x = p$ and $x = q$**
Horizontal line:	$y = c$	$A = c(p - q)$
Sloping straight line:	$y = bx$	$A = \frac{b}{2}(p^2 - q^2)$
Parabola:	$y = ax^2$	$A = \frac{a}{3}(p^3 - q^3)$

Exercise B (Answers Page 138) Use the **difference of x-limits** to find the area under:

a $y = 2$ between $x = 1$ and 5 **b** $y = 5$ between $x = 3$ and 4

Use the **difference of squares** to find the area under:

c $y = 2x$ between $x = 1$ and 5 **d** $y = 5x$ between $x = 3$ and 4

Use the **difference of cubes** to find the area under:

e $y = 2x^2$ between $x = 1$ and 5 **f** $y = 5x^2$ between $x = 3$ and 4

COMBINATIONS

EXAMPLE 3

Suppose we want the area under $y = x + 3$ from $x = 1$ to $x = 2$.

- The area is shown opposite.

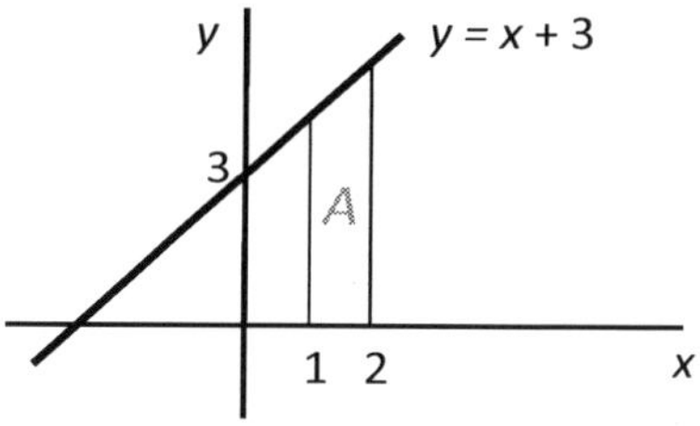

But this area A can be split into two areas B and C by the dashed line in the diagram below.

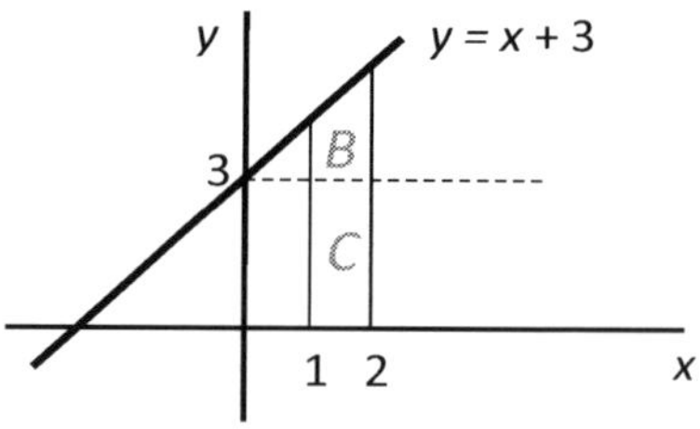

But area B is the same as the area under $y = x$, from $x = 1$ to $x = 2$,

and area C is the same as the area under $y = 3$ from $x = 1$ to $x = 2$.

This shows that we can find the area under $y = x + 3$ by adding the areas under the separate terms: $y = x$ and $y = 3$.

EXAMPLE 4

Now suppose we need the area under $y = 3x^2 + 6x + 7$ from $x = 2$ to $x = 4$.

- A sketch shows that the required area is entirely above the x-axis.

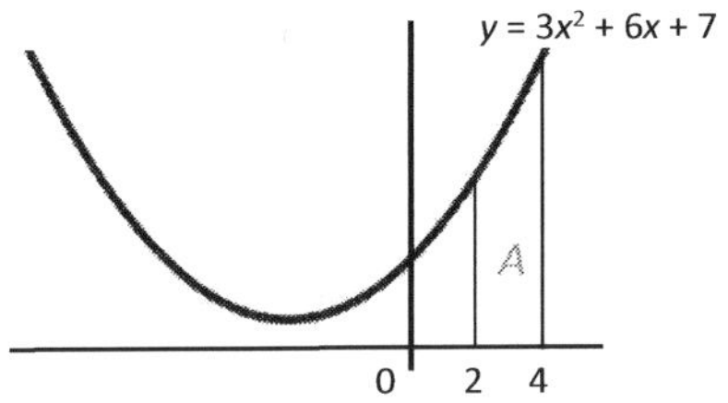

Our equation here has three terms.
But we can **add the individual results**:
for $y = 3x^2$, $y = 6x$ and $y = 7$.

For $y = 3x^2$ we get $\frac{3}{3}(4^3 - 2^3) = 56$.

For $y = 6x$ we get $\frac{6}{2}(4^2 - 2^2) = 36$.

For $y = 7$ we get $7(4 - 2) = 14$.

So the total area is $56 + 36 + 14 =$ **106**.

EXAMPLE 5

Find the area under $y = 12 - 3x^2$ from $x = 2$ to $x = 3$.

- A sketch shows that the curve cuts the x-axis at $x = -2$ and $x = 2$. And therefore the entire area required is below the x-axis (rather than partly above and partly below).

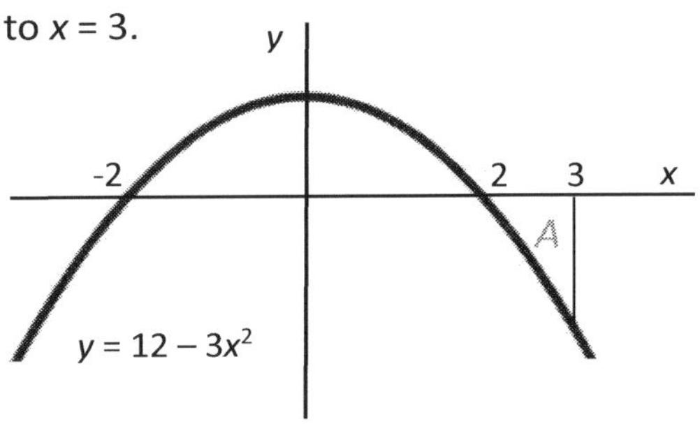

So we can go ahead and use the two x-limits.

The first term is $y = 12$ indicates 12 times a difference of x values,
then second term $-3x^2$ indicates -3 times a third of the difference of cubes.

So that gives $12(3-2) - 3\,\frac{1}{3}(3^3 - 2^3)$

$= 12 - 19$

$= -7.$

So the required area is **7** and the minus indicates it is below the *x*-axis.

EXAMPLE 6

Find the area between $y = x^2 - 4x + 5$, the *x*-axis and $x = 1$ and $x = 4$.

- We see the curve is a parabola, and U-shaped. If the turning point (the minimum point) is above the *x*-axis the curve will not cross the *x*-axis.

  So we can find the *x*-coordinate of the turning point using differentiation (see Chapter 4) and use that to find the *y*-coordinate.

  We get $2x - 4 = 0$, and so $x = 2$.
  And $y = 2^2 - 4\times 2 + 5 = 1$.

  This means the curve is entirely above the *x*-axis:

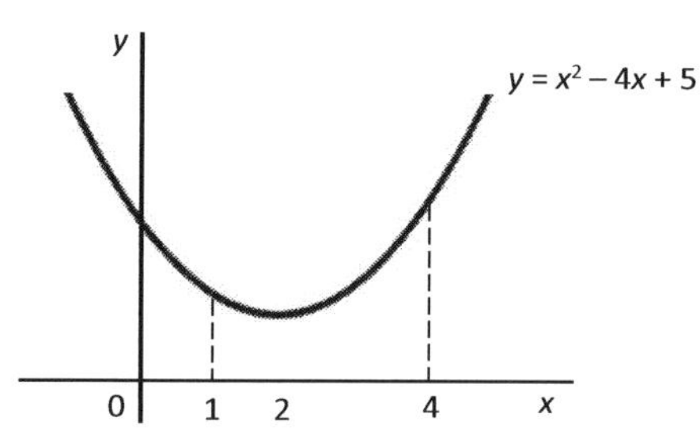

We get $A = \frac{1}{3}(4^3 - 1^3) - \frac{4}{2}(4^2 - 1^2) + 5(4 - 1) = \mathbf{6}.$

<u>**Exercise C**</u> (Answers Page 138) Find the area between the following curves, the *x*-axis and the given limits for *x* (draw a sketch in each case):

a $y = x^2 - 4x$, $x = 0$ to 4 **b** $y = 2x^2 + 5$, $x = 1$ to 3

c $y = x^2 + 1$, $x = -1$ to 2 **d** $y = 3x - x^2$, $x = 0$ to 3

CUBICS ETC.

Now we are in a position to extend to many other areas.

We have noticed a pattern in the formulae for getting areas under $y = a$, $y = ax$, $y = ax^2$.

	Equation	Area in 1st quadrant, from $x = p$ and $x = q$
Horizontal line:	$y = a$	$A = a(p - q)$
Sloping straight line:	$y = ax$	$A = \frac{a}{2}(p^2 - q^2)$
Parabola:	$y = ax^2$	$A = \frac{a}{3}(p^3 - q^3)$
Cubic:	$y = ax^3$	

So we can predict the next line, for $y = ax^3$.
What do you think it will be?

We suspect that for $y = ax^3$, $A = \frac{a}{4}(p^4 - q^4)$.

Recall that for $y = ax$ we had a ratio of areas of 1:1, and for $y = ax^2$ we had a ratio of 2:1, as shown below.

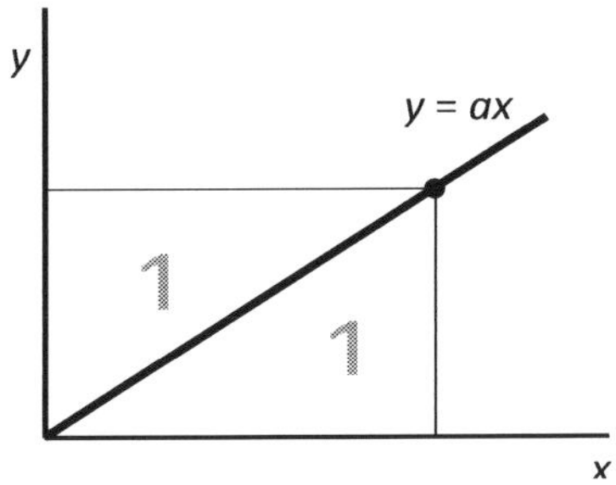

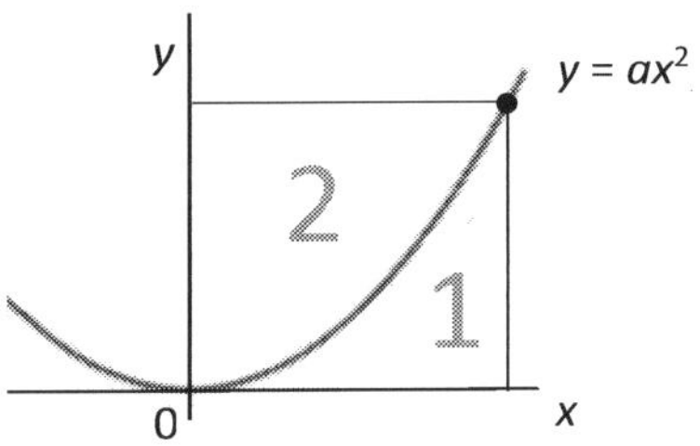

So for $y = ax^3$ we expect a ratio of 3:1 [this is proved on Page 121, 2].

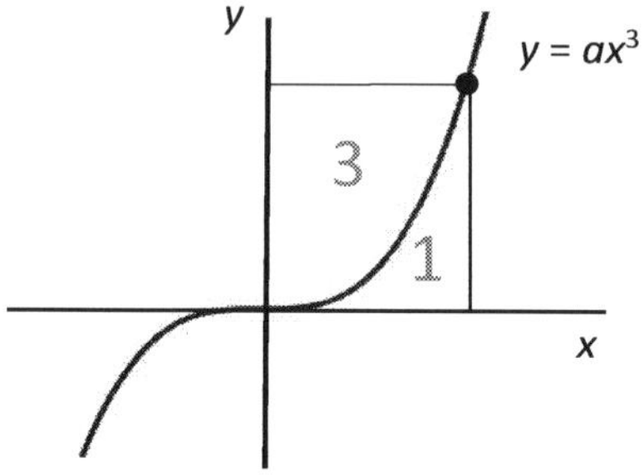

For $y = ax^4$ we have a ratio of 4:1.

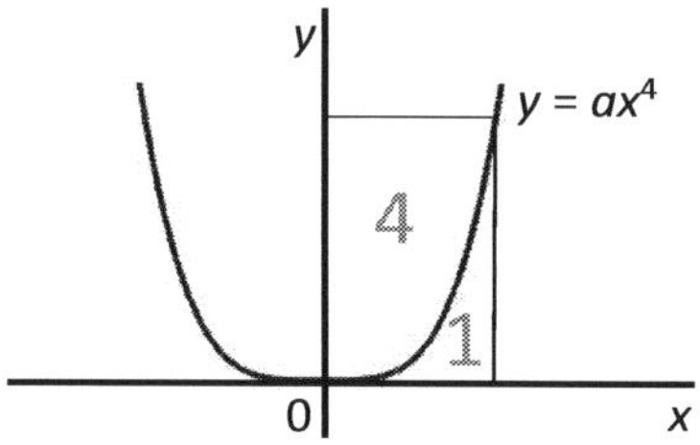

And we can similarly find further results leading us to:

for $\boldsymbol{y = ax^n}$ **the ratio of areas is** $\boldsymbol{n}$**:1**.

[This is proved on Pages 122, 3]

In fact this result is true even if n is a fraction.
So if $n = \frac{2}{3}$ for example, as illustrated.

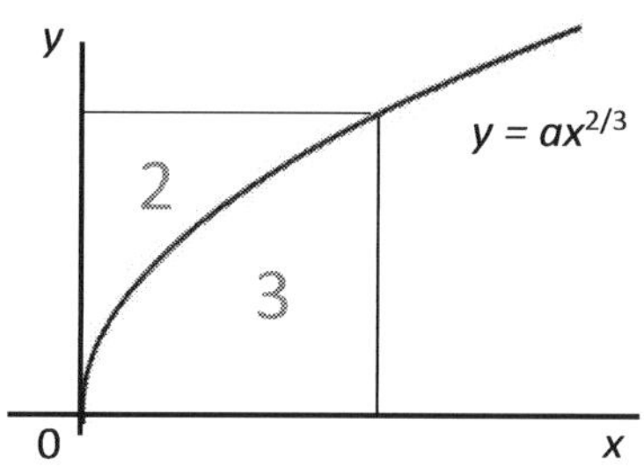

In Chapter 8 we saw that a ratio of gradients was equal to n.

Combining this result with our ratio of areas we can say:

For $y = ax^n$
the ratio of gradients = the ratio of areas = n.

[An alternative proof of this result can be found on Pages 123, 4]

EXAMPLE 7

Find the area under $y = 2x^3$ from x = 1 to x = 2.

- We have $A = \frac{2}{4}(2^4 - 1^4) = \frac{15}{2}$.

<u>**Exercise D**</u> (Answers Page 138) Find the area under:

a $y = x^3$ from $x = 1$ to $x = 4$ **b** $y = 2x^3$ from $x = 0$ to $x = 3$

c $y = x^4$ from $x = 1$ to $x = 2$

d Look at the table on Page 83 and write down a formula that would go in the 3rd column if the Equation in the middle column was $y = ax^n$.

e Find the area shown under $y = x^{2/3}$ from $x = 0$ to $x = 8$. (see the diagram on Page 84)

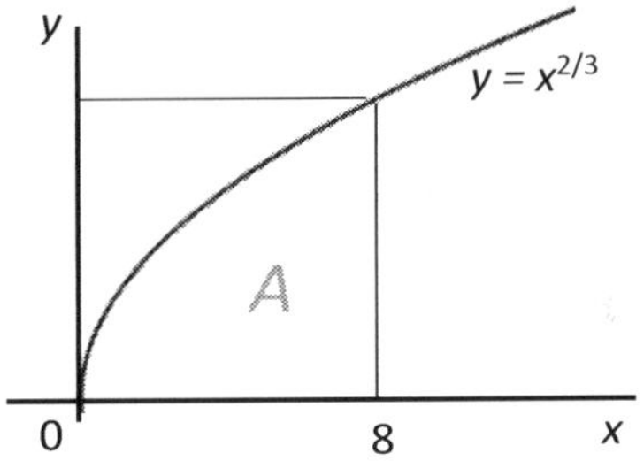

EXAMPLE 8

Find the area under $y = 4x + 2x^3$ from $x = 1$ to $x = 3$.

- We can see here that the area must be entirely above the x-axis as the coefficients and x-limits are all positive.

  We will have a difference of squares for the $4x$ term, and a difference of fourth powers for the x^3 term.

  So $A = \frac{4}{2}(3^2 - 1^2) + \frac{2}{4}(3^4 - 1^4) = 16 + 40 = \mathbf{56}$.

EXAMPLE 9

Find the area under $y = 2x^3 - 3x^2 + 4x + 5$ from $x = 2$ to $x = 4$.

We will assume here that we know the required area is entirely above the x-axis.

- We have four terms and there will be differences of 4th powers, cubes, squares and x-limits.

  So $A = \frac{2}{4}(4^4 - 2^4) + \frac{-3}{3}(4^3 - 2^3) + \frac{4}{2}(4^2 - 2^2) + 5(4 - 2)$

  $= 120 - 56 + 24 + 10 = 98$.

Exercise E (Answers Page 138) Find the area under (you can assume that the areas are entirely above or below the x-axis):

a $y = x^3 + 2x^2$ from $x = 1$ to $x = 3$ **b** $y = 5x - 2x^3$ from $x = 2$ to $x = 3$.

c $y = 7 - 2x + 3x^3$ from $x = 0$ to $x = 3$.

Key Points of Chapter 10

* For: $y = a,$ $A = a(p - q)$

$y = ax,$ $A = \frac{a}{2}(p^2 - q^2)$

$y = ax^2,$ $A = \frac{a}{3}(p^3 - q^3)$

$y = ax^3,$ $A = \frac{a}{4}(p^4 - q^4)$ and so on, leading to the general result:

$y = ax^n,$ $A = \frac{a}{n+1}(p^{n+1} - q^{n+1})$.

* We can use these patterns/formulae to find the area of a strip under a curve between two x-limits (p and q).

* We can also use them in sequence to find areas under curves which are described with two or more terms. But we may need to sketch the graph in this case to avoid areas that are partly above and partly below the x-axis.

AHEAD

We move on to integration now, which is the reverse of differentiation. Then in Chapter 12 we show that the method we have been using in these last two chapters is in fact differentiation in reverse.

CHAPTER 11

Integration

In this Chapter:

*** What integration is**
*** What the 'constant of integration' is and how to find it**
*** The quick way to integrate**
*** The symbol for integration**
*** How to integrate a series of terms**

You are now familiar with the process of differentiating. Integration is simply the reverse of that.

For example, differentiating $y = 5x^4$ gives $\frac{dy}{dx} = 20x^3$.

So integrating $\frac{dy}{dx} = 20x^3$ gives $y = 5x^4$.

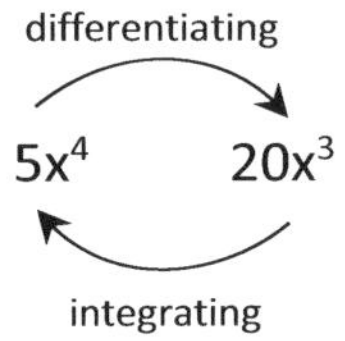

EXAMPLE 1

Which of the following gives $15x^4$ when differentiated:

A $60x^3$ B $15x^4$ C $3x^5$ D $15x^5$

- Differentiating each of these four choices we find that only C gives $15x^4$, so that is the answer.

Note that just as the power is reduced by 1 when we differentiate, when we integrate the power is increased by 1.

<u>**Exercise A**</u> (Answers Page 139)

a Which one of the following gives $\mathbf{6x^2}$ when differentiated:

A $12x$ B $3x^2$ C $2x^3$ D $6x^3$

b Which one of the following gives $\mathbf{8x^3}$ when differentiated:

A $2x^4$ B $2x^3$ C $8x^4$ D $24x^2$

THE CONSTANT OF INTEGRATION

There is a neat formula for doing integration, but first we need to note one thing . . .

Differentiating $y = 5x^4$ gives $\frac{dy}{dx} = 20x^3$ as we said.

But differentiating $y = 5x^4 + 1$ also gives $\frac{dy}{dx} = 20x^3$.

And differentiating $y = 5x^4 + 100$ or $y = 5x^4 - 3$ etc. all give $20x^3$.

So when we integrate we don't know what constant we need to put on the end.

The best we can do is to state that if $\frac{dy}{dx} = 20x^3$ then $y = 5x^4 + c$, where c is some constant whose value we don't know.

We call this the "constant of integration".

EXAMPLE 2

Similarly if $\frac{dy}{dx} = 8x$ then we can say $y = 4x^2 + c$.

We say: **the integral of $8x$ is $4x^2 + c$.**

<u>Exercise B</u> (Answers Page 139) Find y for the following:

a $\dfrac{dy}{dx} = 6x^2$ **b** $\dfrac{dy}{dx} = 8x^3$ **c** $\dfrac{dy}{dx} = 15x^4$

THE PATTERN

In general, the integral of $ax^n = \dfrac{ax^{n+1}}{n+1} + c$

You can see this is correct because if you differentiate

$$\frac{ax^{n+1}}{n+1} + c$$

the $(n+1)$ cancels, the power goes down by 1 and so you get ax^n.

EXAMPLE 3

So to integrate $12x^3$ we get $\dfrac{12x^4}{4} + c = 3x^4 + c$.

The easiest way to think about this is to increase the power by 1, and also divide by the new power.

So for $12x^3$ we increase the power by 1 to get $12x^4$ and also divide 12 by 4 to get $3x^4$.

power increased by 1

Integral of $12x^3 = \dfrac{12x^4}{4} + c$ also divide by the new power

After some practice you will be able to get $3x^4$ in one go.

EXAMPLE 4

Integrate $6x^6$

- We get $\frac{6}{7}x^7 + c$

<u>Exercise C</u> (Answers Page 139) Integrate the following:

a $5x^2$	**b** $20x^3$	**c** $5x^4$	**d** x^3
e x^4	**f** $2x^6$	**g** $\frac{1}{2}x^3$	**h** $\frac{2}{7}x$

INTEGRATION SYMBOL

Just as we used the symbol $\frac{d}{dx}$ for differentiating, we have a symbol for integrating.
We write:

$$\int 20x^3 dx = 5x^4 + c.$$

That is, an elongated 'S' followed by the expression being integrated followed by '*dx*'.
So the expression being integrated goes between '$\int$' and '*dx*'.

[See the note on the significance of this integration symbol on Pages 124, 5]

EXAMPLE 5

Similarly, since $\frac{d}{dx}(x^2) = 2x$, we can write:

- $\int 2x\,dx = x^2 + c.$

EXAMPLE 6

Find $\int 9x^2\, dx$

- We get $\frac{9}{3}x^3 + c = 3x^3 + c$.

We can confirm this is correct by differentiating the answer: we expect to get $9x^2$.

EXAMPLE 7

If $\frac{dy}{dx} = 10x^3$ find y.

- We have to integrate $10x^3$ to get y.
  So $y = \frac{10}{4}x^4 + c = \frac{5}{2}x^4 + c$.

Exercise D (Answers Page 139) Note: until you are confident and accurate you should always check your answer by differentiating it.

Integrate:

a $12x^2$ **b** $12x^3$ **c** $12x^4$ **d** $12x$ **e** 12

Find:

f $\int x^2\, dx$ **g** $\int 15x^2\, dx$ **h** $\int \frac{1}{2}x^3\, dx$ **i** $\int 3x^7\, dx$ **j** $\int 2\, dx$

k $\int a\, dx$ **l** $\int ax\, dx$ **m** $\int ax^2\, dx$

n If $\frac{dy}{dx} = 3x^2$ find y **o** If $\frac{dy}{dx} = 9x^8$ find y **p** If $\frac{dy}{dx} = 14x^6$ find y

q If $\frac{dy}{dx} = 4x^5$ find y **r** If $\frac{dy}{dx} = -17x$ find y

INTEGRATING TERM BY TERM

Just as we can differentiate an expression consisting of several terms by differentiating each term individually, so we can integrate term by term too.

EXAMPLE 8

Find $\int (12x^3 + 4x^2 + 3x + 2)dx$

- We integrate each of the four terms.

$$\int (12x^3 + 4x^2 + 3x + 2)dx$$
$$= 3x^4 + \tfrac{4}{3}x^3 + \tfrac{3}{2}x^2 + 2x + c.$$

Exercise E (Answers Page 139) Find:

a $\int (6x^2 + 8x + 10)dx$ **b** $\int (3 - 6x)dx$ **c** $\int (x^2 + x + 1)dx$

d $\int (5x - 4)dx$ **e** Given that $\frac{dy}{dx} = 5x^4 - 18x^2$ find y.

FINDING THE CONSTANT

Sometimes it is possible to find the value of the constant of integration, as the following examples illustrate.

EXAMPLE 9

Given that $\frac{dy}{dx} = 3x^2 - 4x + 1$ and that $y = 1$ when $x = 0$, find y.

- We know we can integrate $\frac{dy}{dx} = 3x^2 - 4x + 1$.

  We get $y = x^3 - 2x^2 + x + c$.

We get a constant of course. But we are also given the values, $y = 1$ when $x = 0$, and we can use these to find the value of c.

We substitute the values $y = 1$, $x = 0$ into $y = x^3 - 2x^2 + x + c$ and get:

$1 = 0 + c$, which gives $c = 1$, and so we can put 1 instead of c:

$\mathbf{y = x^3 - 2x^2 + x + 1}$.

EXAMPLE 10

If $\frac{dy}{dx} = 10x + 5$ find y, given that $y = 2$ when $x = 1$.

- From $\frac{dy}{dx} = 10x + 5$ we get $y = 5x^2 + 5x + c$.

  And substituting $y = 2$, $x = 1$, we get:

  $2 = 5 + 5 + c$

  So $c = -8$ and $\mathbf{y = 5x^2 + 5x - 8}$.

Exercise F (Answers Page 139)

a If $\frac{dy}{dx} = 2x + 3$ find y, given that $y = 3$ when $x = 0$.

b If $\frac{dy}{dx} = 3x^2 + 4x - 5$ find y, given that $y = 1$ when $x = 1$.

c If $\frac{dy}{dx} = 2 - 3x^2$ find y, given that $y = -2$ when $x = -1$.

Key Points of Chapter 11

* Integration is the reverse of differentiation.

* We integrate with the formula: $\int ax^n \, dx = \dfrac{ax^{n+1}}{n+1} + c$.

* We get a 'constant of integration' when we integrate.

* We can integrate a series of terms by integrating each term individually.

* We can find the value of the constant of integration if we are given additional information.

AHEAD

Next we show that integration will give us areas under curves, and is actually the same as we were doing in earlier chapters. Then we move on in Chapter 13 to some applications of both differentiation and integration.

CHAPTER 12

Integration and Area

In this Chapter:

*** How integration and areas are connected**

*** How we can use the integration notation to describe areas**

INTEGRATION GIVES AREAS

When we find the area under $y = x^2$ between $x = 1$ and $x = 4$ we find the value of

$\frac{1}{3}(4^3 - 1^3)$ which is the difference between $\frac{x^3}{3}$ when $x = 4$ and $\frac{x^3}{3}$ when $x = 1$.

But integrating x^2 also gives $\frac{x^3}{3}$.

This indicates we can find areas by using integration. In this example we would integrate x^2 to get $\frac{x^3}{3}$ and then substitute $x = 4$ and $x = 1$ and subtract. This gives the area: $\frac{1}{3}(4^3 - 1^3)$.

NOTATION

The area under $y = x^2$ between $x = 1$ and $x = 4$ is usually written like this:

$$A = \int_1^4 x^2 dx$$

Note that the limits (the 1 and 4) are written immediately after the integration symbol. The greater limit is always on top.

So $\int_1^4 x^2 dx$ means the area between $y = x^2$, the x-axis, and the lines $x = 1$ and $x = 4$, as shown below.

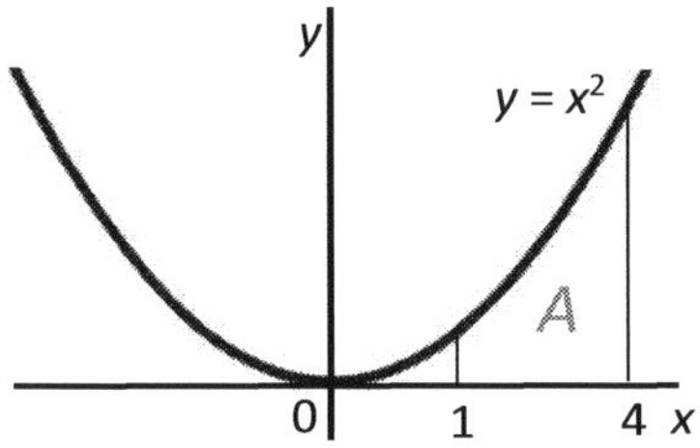

$$A = \int_1^4 x^2 dx$$

And it is evaluated by integrating y, substituting the limits (the upper one first) and subtracting the results.

EXAMPLE 1

Similarly $\int_3^5 x^2 dx$ means the area under $y = x^2$ between $x = 3$ and $x = 5$.

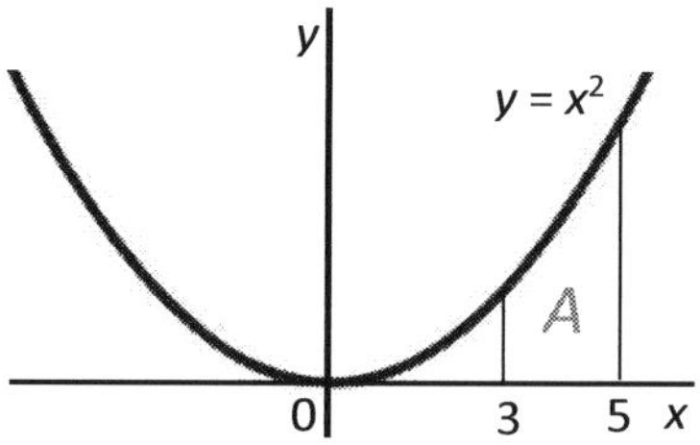

- It's normally worked out like this:

$$\int_3^5 x^2 dx = \tfrac{1}{3}\left[x^3\right]_3^5 = \tfrac{1}{3}[5^3 - 3^3] = \mathbf{\tfrac{98}{3}}.$$

Note: we integrate x^2 to get $\frac{1}{3}x^3$, substitute the upper and lower limits, 5 and 3, for x, then subtract.

Square brackets are used to contain the integrated expression, and the limits go at the right-hand end of them.

Exercise A (Answers Page 140)

a Sketch the area represented by $\int_2^4 x^2 dx$

b Find the area represented by $\int_2^4 x^2 dx$

c Write the integral for the area shown:

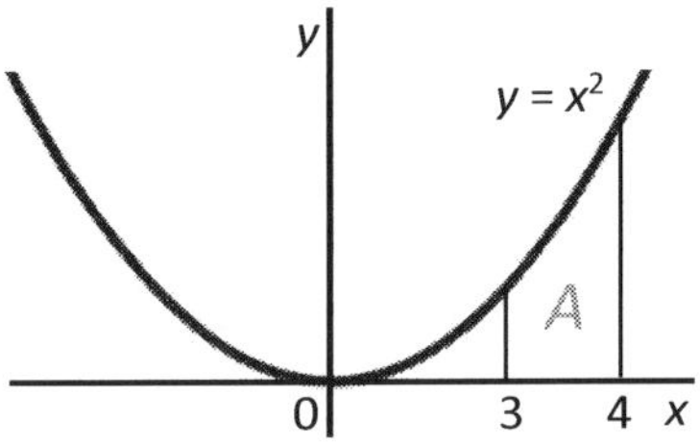

d Find the value of the integral: $\int_2^5 3x^2\, dx$

e At the end of Chapter 10 we arrived at the general formula:

If $y = ax^n$, then $A = \frac{a}{n+1}(p^{n+1} - q^{n+1})$, where A is the area under $y = ax^n$, between $x = p$ and $x = q$.

Show that integration also gives areas, by showing that $\int_q^p ax^n dx = \frac{a}{n+1}(p^{n+1} - q^{n+1})$.

INTEGRATING SEVERAL TERMS

Just as we can integrate an expression with several terms by summing the separate integrals, we can get the area under a curve described by several terms by treating the terms individually.

So for example, $\int_1^2 (3x^2 - 6x)dx = \int_1^2 3x^2\, dx + \int_1^2 (-6x)dx$.

EXAMPLE 2

Find $\int_1^2 (3x^2 - 6x)dx$

- $\int_1^2 (3x^2 - 6x)dx = \left[x^3 - 3x^2\right]_1^2$

$= (2^3 - 3\times2^2) - (1^3 - 3\times1^2)$
$= (-4) - (-2)$
$= -2.$

Note that our previous method of getting difference of squares, difference of cubes etc. will always give the same answer. In this example we would proceed like this:

- $\int_1^2 (3x^2 - 6x)dx = \frac{3}{3}(2^3 - 1^3) - \frac{6}{2}(2^2 - 1^2)$

$= 7 - 9$

$= -2.$

EXAMPLE 3

Find $\int_{-2}^{0} (\frac{1}{2}x^2 - 2x + 4)\ dx$

- $\int_{-2}^{0} (\frac{1}{2}x^2 - 2x + 4)\ dx = \left[\frac{1}{6}x^3 - x^2 + 4x\right]_{-2}^{0} = [0 - 0 + 0] - [\frac{1}{6}(-8) - 4 - 8] = \mathbf{13\frac{1}{3}}.$

Again, for comparison, if we find differences of cubes, squares and *x*-values we get:

$\frac{1}{6}[0+8] - \frac{2}{2}\ [0-4] + 4[0+2] = \mathbf{13\frac{1}{3}}.$

You can use whichever method you prefer.

Exercise B (Answers Page 140)

a Find $\int_1^2 (6x^2 - 2x + 1)dx$

b Find the area under $y = 2x^2$ between $x = 3$, $x = 4$ and the *x*-axis.

c If the curve in **b** is raised 2 units what will be the new area under it between $x = 3$, $x = 4$?

Key Points of Chapter 12

* Integration gives areas under curves.

* We can integrate to get areas, using $A = \int_q^p y\,dx$

where y is an expression in terms of x, and p and q are the limits of integration.

AHEAD

In the next chapter we see how differentiation and integration can be used to study motion. In the final chapter we see another way that differentiation and integration are connected in Differential Equations.

CHAPTER 13

Motion

In this Chapter:

*** How a formula can give positions, speeds and accelerations**

*** How these formulae are related through differentiation and integration**

In this chapter we are going to consider motion in a straight line where the motion is described by some formula involving time.

The units will be metres and seconds.

DISPLACEMENT

A distance is just a length: a number of meters, feet etc.

We use the word **displacement** here instead of distance because displacement includes direction: positive or negative in our case.

We will assume here that displacements are positive to the right, and therefore negative to the left of some fixed point *O*.

So if an object has a displacement of 3 meters it means it is moved a distance 3 meters to the right of *O*.

Similarly a displacement of -2 meters means 2 meters to the left of *O*.

We use the letter *s* for displacement.

EXAMPLE 1

So we might have the formula:

$$s = 5t^2 + 4t$$

for the position of a particle moving in a straight line.

For each value we choose for t we can find the corresponding value for s.

At $t = 0$ the formula gives us $s = 0$ metres.

Similarly we find that after 1 second, $s = 5 + 4 = 9$ m.
Showing that the particle has travelled 9 m in the first second.
And if $t = 2$ we get $s = 28$ m.

We can use a diagram to record the results:

We see the speed of the particle appears to be increasing, since it is covering a greater distance in each following second.

EXAMPLE 2

For another particle governed by $\mathbf{s = 2t^2 - 5t}$, if we put $t = 0, 1, 2, 3, 4$ we get:

$t =$	0	1	2	3	4
$s =$	0	-3	-2	3	12

We see that when $t = 0$ we get $s = 0$, so we know the particle starts at the origin.

The negative values indicate that the particle is to the left of the starting point when $t = 1$ and $t = 2$:

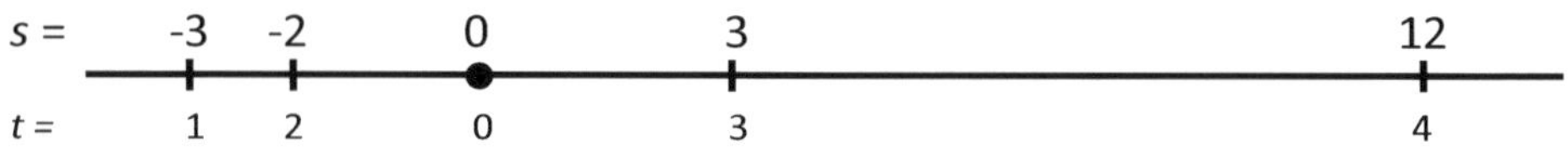

We find the particle appears to be going backwards at first and then starts to move to the right.

Exercise A (Answers Page 140)

Given $s = 5t - t^3$ find the displacement after 0, 1, 2, 3 seconds.

Briefly describe this part of the motion.

VELOCITY

Just as the word 'displacement' is used where direction needs to be included, the word **velocity** is used (rather than 'speed') when direction needs to be included.

Here, our velocities will increase to the right, so a positive velocity means a movement to the right and a negative velocity means it is to the left.

A velocity of **7 meters per second** means that the particle's displacement is increasing by **7 meters** every second.
We write 'meters per second' as 'm/s'.

We use the letter v for velocity.

GETTING VELOCITY FROM DISPLACEMENT

In fact, if we have a formula for displacement we can get a formula for velocity from it, because velocity is the rate at which displacement is increasing.

That is to say: $v = \frac{ds}{dt}$.

So given $\boldsymbol{s = 5t^2 + 4t}$ we can say that velocity, $v = \frac{ds}{dt} = \boldsymbol{10t + 4}$.

differentiating

$$s \longrightarrow v = \frac{ds}{dt}$$

So when we have an expression for displacement we can use calculus to get an expression for velocity.

EXAMPLE 3

The displacement of a particle from an origin at time t is given by:

$$\boldsymbol{s = 3t^2 - 4t^4}$$

Find an expression for the velocity, and find the velocity after 3 seconds.

- Differentiating we get, $\boldsymbol{v = 6t - 16t^3}$.

- And when $t = 3$, we find $v = 6\times3 - 16\times3^3 = 18 - 432 = -414$ m/s.

EXAMPLE 4

Given that the motion of a particle is given by $\boldsymbol{s = 30 + 4t - 5t^2}$ determine:
a) the initial velocity
b) when the particle is at rest.

- a) the 'initial velocity' means the velocity when $t = 0$.
  So we get the velocity formula: $\boldsymbol{v = 4 - 10t}$, and put $t = 0$ to get $\boldsymbol{v = 4}$ **m/s**.

- b) 'at rest' means the velocity is zero. So we equate v to zero in our velocity formula: $v = 4 - 10t$:

  We get $4 - 10t = 0$

  and so $t = 0.4$ sec.

  The particle is at rest after **0.4 seconds**.

Exercise B (Answers Page 140)

a Given $s = 5t^2 - 3t$ find the velocity after 3 seconds.

Also find when the particle is at rest.

b If $s = 30 - t^3$ find v when $t = 2$.

c If $s = 3t^2 - 12t + 5$ find the initial velocity and when the particle is at rest.

ACCELERATION

Acceleration is also a rate of increase: the rate at which velocity is increasing. Like displacement and velocity, it also includes direction.

So for example an acceleration of **5 meters per second per second** means that the velocity is increasing by **5 meters per second** in every second.
We write 'meters per second per second' as: 'm/s²'.

And since acceleration, a, is the rate of increase of velocity we have $a = \frac{dv}{dt}$.

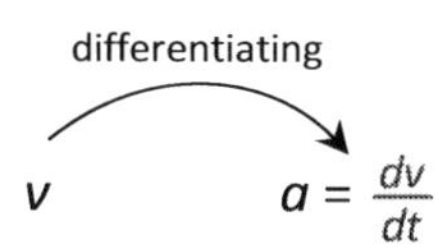

This means that if we have an expression in terms of time for velocity, we can differentiate it to get an expression for acceleration.

EXAMPLE 5

If $\boldsymbol{v = 3t^2 - 5t}$ then $a = \frac{dv}{dt} = 6t - 5$.

EXAMPLE 6

Given $\boldsymbol{v = 10t - 2t^2}$ find the acceleration when $t = 2$.

- $a = \frac{dv}{dt} = 10 - 4t$ and so when $t = 2$, $a = 10 - 8 = 2$ m/s^2.

DISPLACEMENT, VELOCITY, ACCELERATION

Since differentiating displacement gives velocity, and differentiating velocity gives acceleration, it means we can get acceleration by differentiating the displacement expression twice.

$$s \xrightarrow{\text{differentiating}} v = \frac{ds}{dt} \xrightarrow{\text{differentiating}} a = \frac{d^2s}{dt^2}$$

You may recall from Chapter 5 that $\frac{d^2s}{dt^2}$ is the second derivative.

EXAMPLE 7

Given $\boldsymbol{s = t^3 - 2t^2 + 4t}$ find the velocity and acceleration after 5 seconds.

- Differentiating we get $\boldsymbol{v = 3t^2 - 4t + 4}$, and putting $t = 5$, $\boldsymbol{v = 59}$ **m/s**.
- Differentiating $v = 3t^2 - 4t + 4$, we get $\boldsymbol{a = 6t - 4}$, and putting $t = 5$, $\boldsymbol{a = 26}$ **m/s²**.

Exercise C (Answers Page 140)

a Given $s = t^3 - 4t^2 - 5t - 6$ find expressions for velocity and acceleration.

b If $s = t^4 - 3t^2 + 1$ find the displacement, velocity and acceleration after 2 seconds.

c Find an expression for acceleration given that $v = 12t^2 + 7t + 10$.
What is the initial velocity and initial acceleration?

d The height of a projectile which is fired vertically upwards is given by $h = 20t - 5t^2$.
(Here we take the upwards direction as positive.)
What is its initial velocity?
When is the velocity zero?
What is the greatest height reached?

USING INTEGRATION

Because integration is the reverse process of differentiation:
we can *get velocity by integrating acceleration*,
and *get displacement by integrating velocity*, as shown below.

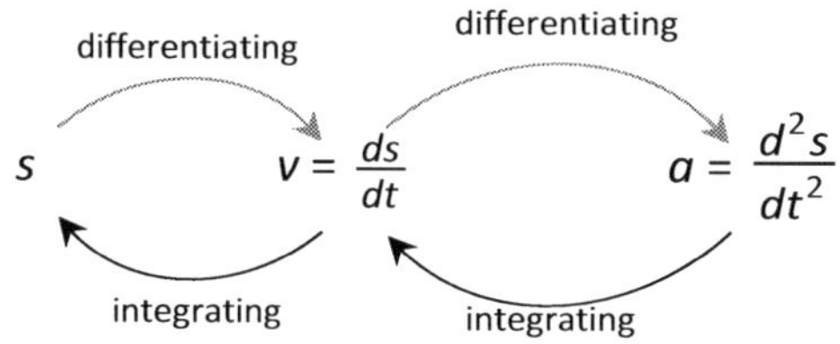

An example will make this clear.

EXAMPLE 8

Given $\boldsymbol{v} = \mathbf{3}\boldsymbol{t}^{\mathbf{2}} + \mathbf{6}\boldsymbol{t} + \mathbf{1}$, find expressions for displacement and acceleration, given that $s = 0$ when $t = 0$.

Notice velocity is in the middle between displacement and acceleration in the above diagram.

- We can differentiate v to get a, as before: $\boldsymbol{a} = \mathbf{6}\boldsymbol{t} + \mathbf{6}$.
  And we can integrate v to get s: $s = t^3 + 3t^2 + t + c$.

  Since we also know that $s = 0$ when $t = 0$, we can find the constant of integration, c.

  We get $c = 0$ and so $\boldsymbol{s} = \boldsymbol{t}^{\mathbf{3}} + \mathbf{3}\boldsymbol{t}^{\mathbf{2}} + \boldsymbol{t}$.

EXAMPLE 9

Find expressions for velocity and displacement given $\boldsymbol{a} = \mathbf{6}\boldsymbol{t} + \mathbf{4}$ and that when $t = 0$, $s = 10$ and $v = 5$.

- Since $a = 6t + 4$ we can integrate to get velocity:
  $v = 3t^2 + 4t + \mathrm{c}$
  And since we are given that when $t = 0$, $v = 5$ we can find c:

$5 = 0 + c$.

So $c = 5$ and $\boldsymbol{v = 3t^2 + 4t + 5}$.

- For displacement we integrate velocity. We get:

  $s = t^3 + 2t^2 + 5t + c$.

  And since we know that when $t = 0$, $s = 10$ we can find the new c:

  $10 = 0 + c$.

  We find $c = 10$ here and so $\boldsymbol{s = t^3 + 2t^2 + 5t + 10}$.

Exercise D (Answers Page 141)

a Given $v = 6t^2 + 8t - 1$, find expressions for displacement and acceleration.

b Find expressions for v and s given that $a = 12t$ and that when $t = 0$, $v = 7$ and $s = 0$.

c Given $v = 3t^2 - 24t$ find the value of t for which $a = 0$.
Given that $s = 20$ when $t = 0$ find s when $t = 2$.

Key Points of Chapter 13

* We can differentiate displacement, s, to get velocity v, and differentiate velocity to get acceleration, a.

* In reverse we can integrate acceleration to get velocity and integrate velocity to get displacement.

* In this way we can make predictions about these quantities from just one initial fact about either displacement or velocity or acceleration.

AHEAD

In the final chapter we see another way that differentiation and integration are connected in solving Differential Equations.

CHAPTER 14

Differential Equations

In this Chapter:
*** What a differential equation is**
*** How to construct a differential equation**
*** How to use a simple pattern to construct the equation**
*** How to reverse that pattern to get a solution to differential equations**

WHAT IS A DIFFERENTIAL EQUATION?

A differential equation is an equation that includes $\frac{dy}{dx}$ or higher order derivatives like $\frac{d^2y}{dx^2}$.

So we are actually familiar with very simple differential equations, as we have been solving equations like $\frac{dy}{dx} = 4x$.

However, the term "differential equation" usually refers to more elaborate equations like:

$3y + 4\frac{dy}{dx} = 5x + 6$ a first order differential equation, or

$3y + 4\frac{dy}{dx} + 5\frac{d^2y}{dx^2} = 6x^2 + 5x + 1$ a second order differential equation.

There are many situations in which connected variables increase or decrease at certain rates and this can give rise to equations like the ones above, so their solution is very important.

The variety of differential equations that can occur is extremely wide.

For brevity here we will use y' (pronounced "y dashed") instead of $\frac{dy}{dx}$.

So, for example, if $y = 3x^2$, then $y' = 6x$.

CONSTRUCTING A DIFFERENTIAL EQUATION

Suppose we have $\mathbf{y = 2x + 3}$.

Then we could work out say, $\mathbf{5y + 7y'}$.

We get: $\mathbf{5y} = 5(2x + 3) = 10x + 15,$ (1)

and $\mathbf{7y'} = 7 \times 2 = 14$, since $y' = 2$. (2)

So $\mathbf{5y + 7y'} = 10x + 15 + 14 = \mathbf{10x + 29}$.

We get the differential equation $\mathbf{5y + 7y' = 10x + 29}$

and $\mathbf{y = 2x + 3}$ is a solution to it.

It is a solution because if $y = 2x + 3$ then $5y + 7y' = 10x + 29$.

We are going to see how to get the $10x + 29$ above using a simple pattern.
Then we will see that that pattern can be used in reverse to give a solution to the differential equation.

Using the Pattern

EXAMPLE 1

Again, suppose we want to know what $\mathbf{5y + 7y'}$ is when $\mathbf{y = 2x + 3}$.

- Arrange the coefficients like this:

5	**7**	differential expression
2	**3**	*y*
10x +	**29**	result

You can see that the **10** in **10***x* + 29 came from multiplying vertically: 5 × 2 = **10** (see line (1) above).

And the **29** came from cross-multiplying and adding: 5 × 3 + 7 × 2 = **29** (see lines (1) and (2) above).

So the pattern we are looking for is as shown below:

5 7

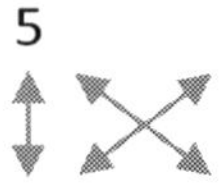

2 3

10 29

That is, we multiply vertically on the left to get 10, and then crosswise and add to get 29. This gives us the 10*x* + 29.

Algebraically

This pattern is easily confirmed algebraically as follows:
If we want to know $Py + Qy'$ when $y = ax + b$, we note that $y' = a$, so:

$Py + Qy' = P(ax + b) + Qa = Pax + (Pb+Qa)$.

And if we set out as we did before then:

P	Q	differential expression
a	b	y
Pax	$Pb+Qa$	result

which is what the vertical/crosswise pattern gives us.

EXAMPLE 2

Find **$3y + 5y'$** when **$y = 8x - 2$**.

- We can do it the long way or use the pattern:

3 5

8 -2

24 **34** where 3×8 = **24**, and 3(-2) + 5×8 = -6 + 40 = **24**.

So, when $y = 8x - 2$, **$3y + 5y' = 24x + 34$**.

<u>**Exercise A**</u> (Answers Page 141)

a Find $2y + 3y'$ when $y = 4x + 5$

b Find $y + 6y'$ when $y = 2x - 3$

c Find $3y - y'$ when $y = 7x + 1$

d Find $4y + y'$ when $y = 5x$

e Show that $y = 4x - 3$ is a solution to $5y + 3y' = 20x - 3$

IN REVERSE

We know from Example 1 that if $y = 2x + 3$ then $5y + 7y' = 10x + 29$.

EXAMPLE 3

We want to ask: **given $5y + 7y' = 10x + 29$, what is y?**

5	7	differential expression

a	***b***	*y*
10	29	result

That is to say, given the differential expression and the result, can we find y? Can we find the values of a and b above?

You might like to try this yourself and see if you can use that pattern in reverse to get y.

We do it like this . . .

Vertically we know that $5 \times a = 10$. So ***a*** must be **2**.

We now have:

5 7

2 ***b***

10 29

Now look at the crosswise step. We know it must come to 29, and we know one part of the cross is $7 \times 2 = 14$.

So the other part of the cross must be 15 (because $29 - 14 = 15$)

b must therefore be **3** because only 5×3 gives 15.

5 7

2 **3**

10 29

So we have $\boldsymbol{y = 2x + 3}$.

Let's take another example.

EXAMPLE 4

Find a solution to: $\boldsymbol{3y + 6y' = 12x + 39}$.

- We start with:

  3 6

  a ***b***

  12 39

  We see $\boldsymbol{a}$ must be **4**, as $3 \times 4 = 12$.

  Now with $a = 4$ we have:

  3 6

  4 ***b***

  12 39

  One cross product is $6 \times 4 = 24$,
  and that leaves 15, as $39 - 24 = 15$.
  So the other cross product is $3 \times b = 15$ and therefore $\boldsymbol{b = 5}$.

  So $\boldsymbol{y = 4x + 5}$.

With a bit of practice you can write the answer straight down.

Exercise B (Answers Page 141) Find a solution to the following differential equations.

a $2y + 3y' = 10x + 17$ **b** $2y + y' = 2x + 7$

c $11y + 3y' = 33x + 31$ **d** $3y + y' = 9x + 6$

e $3y + y' = 3x + 10$ **f** $y + y' = x + 2$

g $3y + y' = 15x + 2$ **h** $2y + 3y' = 6x + 10$

The solutions shown here are not the full solutions. In each case there is another term which needs to be added to the answer to give the complete solution. That step will be covered later.

However the method shown here can be extended to second order differential equations and beyond.

Key Points of Chapter 14

* A Differential Equation is an equation that contains $\frac{dy}{dx}$ or higher order derivatives like $\frac{d^2y}{dx^2}$.

* We can create and find solutions to Differential Equations using a vertical/crosswise pattern.

NOTES

Here are given some results and notes referred to in the main text.

Chapter 3, Page 21

Limits

There is in fact something to consider here because if we go back to the last line of our proof from Pages 14-15 :

$$\text{gradient of PQ} = \frac{(p-q)(p+q)}{p-q} = p+q$$

we cancelled $\frac{p-q}{p-q}$ to get 1.

But we cannot justify cancelling like that when $p = q$, because in that case we would have $\frac{0}{0}$, which cannot be found.

And in fact we have just put $p = q$ to get the gradient of a tangent.

However we can avoid this problem by using limits.

We can say that:

the limit of $\frac{(p-q)(p+q)}{p-q}$ as q approaches p is $2p$,

because as q approaches the value p, $\frac{p-q}{p-q}$ will approach equality and $(p + q)$ will approach $2p$.

Ultimately (in the limit) $\frac{p-q}{p-q}$ will equal 1 and $(p + q)$ will equal $2p$.

This same reasoning applies in later work where we will be looking at other equations but we will not discuss this detail again.

Chapter 6, Page 45

[3] DIFFERENTIATING SEVERAL TERMS

In Chapter 2, Exercise D, Qu. **e**, (see answer on Page 128) we looked at $y = ax^2 + bx + c$ and we had:

$$\text{secant gradient} = \frac{(ap^2 + bp + c) - (aq^2 + bq + c)}{p - q}$$

and we saw that we could cancel c, and we separated out the squared terms to get:

$$\frac{(ap^2 - aq^2) + (bp - bq)}{p - q}.$$

We then saw that $(p - q)$ cancels with both the squared terms and also with the others:

$\dfrac{a(p^2 - q^2) + b(p - q)}{p - q}$ here we have taken a and b out as factors of the brackets

$= \dfrac{a(p-q)(p+q) + b(p - q)}{p - q}$ here we have replaced $p^2 - q^2$ with $(p - q)(p + q)$

$= \dfrac{(p-q)[a(p+q) + b]}{p - q}$ here we have taken $(p - q)$ out as a common factor

$= \boldsymbol{a(p + q) + b}$ here we have cancelled $(p - q)$.

This cancelling will always occur because $(p - q)$ divides into $(p - q)$, $(p^2 - q^2)$, $(p^3 - q^3)$ and so on.

This tells us that we can differentiate equations like $y = 5x^3 + 4x^2 - 7x + 3$ **term by term**.

That is, we differentiate $5x^3$ then $4x^2$ then $-7x$.

And we know the 3 cancels out (like c in the above algebra).

Chapter 6 – Page 51

LAWS OF INDICES – a summary for reference purposes

The Index is the power on a number, e.g. in 3^5 the index of 3 is 5.

1) $a^0 = 1$ E.g. $2^0 = 1$.

Any number (except 0) whose index is 0 is always equal to 1.

2) $a^{-m} = \dfrac{1}{a^m}$ E.g. $2^{-3} = \dfrac{1}{2^3} = \dfrac{1}{8}$.

3) $a^m \times a^n = a^{m+n}$ E.g. $x^3 \times x^4 = x^7$

4) $a^m \div a^n = a^{m-n}$ E.g. $x^8 \div x^5 = x^3$

5) $(a^m)^n = a^{mn}$ E.g. $(x^2)^5 = x^{10}$

6) $a^{m/n} = \sqrt[n]{a^m} = \left(\sqrt[n]{a}\right)^m$ E.g. $8^{2/3} = \left(\sqrt[3]{8}\right)^2 = 4$

Chapter 8, Page 62

RATIO OF GRADIENTS = *n*

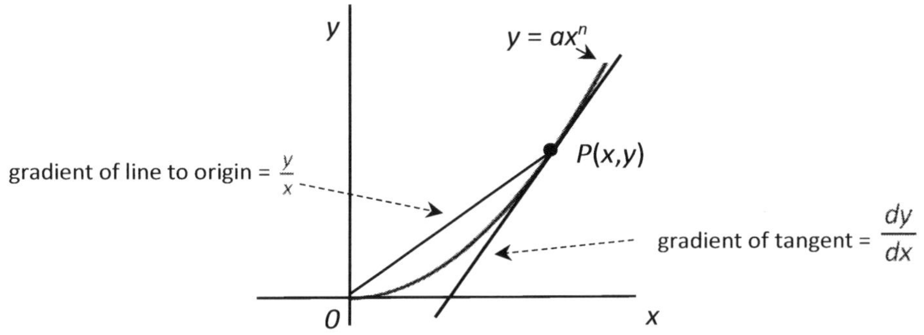

We need to show that $\dfrac{dy/dx}{y/x} = n$.

NOTES

Proof

Starting with $y = ax^n$, the gradient of the tangent is $\frac{dy}{dx} = anx^{n-1}$.

And $\frac{y}{x} = \frac{ax^n}{x} = ax^{n-1}$.

Therefore $\frac{dy/dx}{y/x} = \frac{anx^{n-1}}{ax^{n-1}} = n$.

This can also be proved with the secant-tangent method.

<u>Chapter 9, Page 68</u>

Proof that the ratio of areas for $y = ax^2$ is 2:1.

That is that $\frac{B}{A} = 2$, where A is the value of the area between the curve, the x-axis and the lines $x = p$ and $x = q$. And B is the value of the corresponding area.

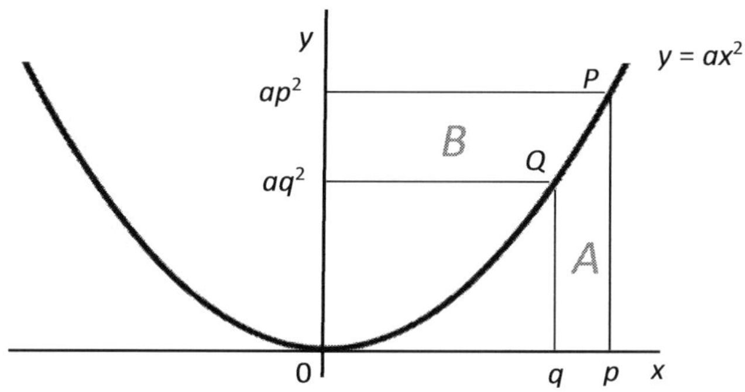

The proof follows similar lines to the proof for the gradient of a curve shown in Chapter 3.

We choose two points P and Q on the curve.

The x-coordinates of P and Q are p and q.
And the y-coordinates are therefore ap^2 and aq^2, as shown.

We are going to find the ratio of the areas B:A.

And we will do that as we did for the gradients by letting *Q* approach *P* along the curve.

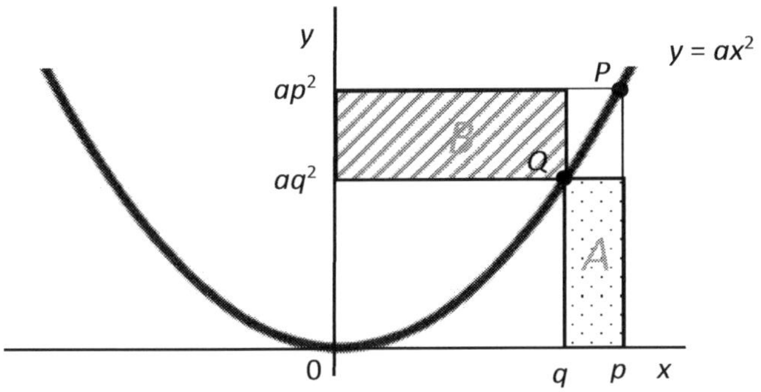

We will approximate area *B* with the rectangle shown shaded diagonally.
Similarly area *A* is approximated by the dotted rectangle.

The area of the upper rectangle is *base* × *height* = $q \times (p^2 - q^2)$

The area of the rectangle on the right is *base* × *height* = $(p - q) \times q^2$.

The ratio of these areas is $\frac{B}{A} = \frac{q(p^2 - q^2)}{(p - q)q^2}$

And remembering that $(p^2 - q^2) = (p + q)(p - q)$ we find we can do some cancelling and we get $\frac{p+q}{q}$.

Now as *Q* approaches *P*, the value of *q* approaches the value of *p*.

When $q = p$ we get the ratio of areas as $\frac{2p}{p}$, which is 2.

What exactly have we proved here?
We have proved that as the horizontal and vertical strips get narrower and narrower the ratio of the areas approaches 2:1, and ultimately equals 2:1.

Note that the ratio of areas is independent of *x*, which means it applies to all such pairs of strips.

NOTES

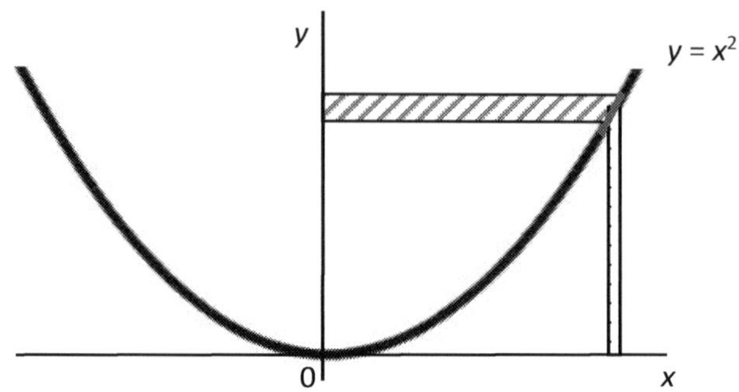

Ratio of areas approaches 2:1

Now we can suppose that any area to the left of $y = x^2$ and bounded by two horizontal lines and the y-axis (like area B) can be split into a large number of strips as shown below.

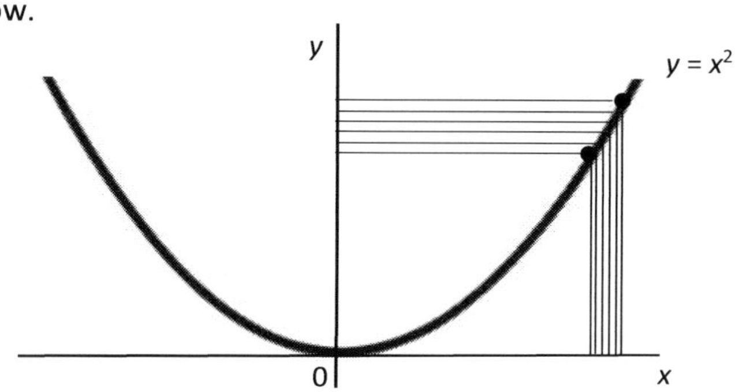

And since the ratio of areas of two corresponding strips (one horizontal and one vertical) is ultimately 2:1, therefore the ratio of the whole area consisting of a collection of strips on the left to the corresponding area on the right will also be 2:1.

Therefore also the ratio of areas from the origin to corresponding boundaries will have the ratio 2:1.

This proves that any area B in $y = x^2$ is double the corresponding area A.

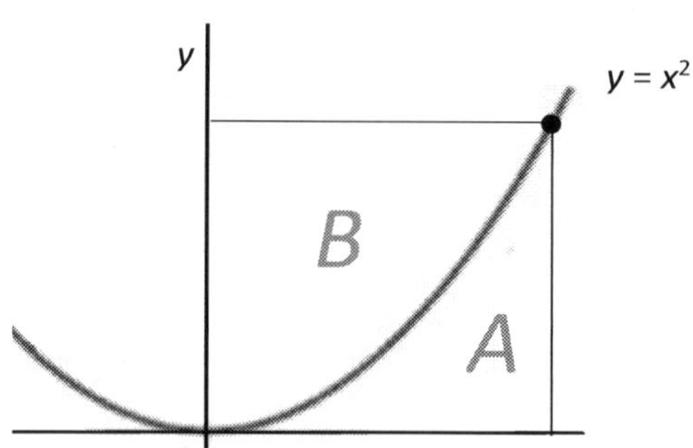

Area B is double area A

And that the area of any horizontal strip is double the area of the corresponding vertical strip.

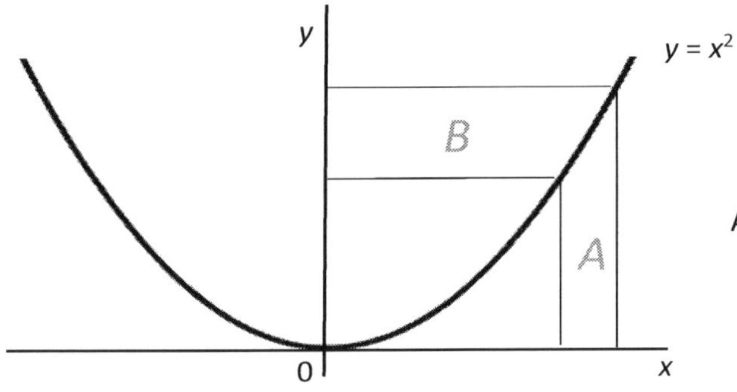

Area B is double area A

Chapter 10, Page 84

To show that for $y = ax^3$ the ratio of areas is 3:1 we follow a similar argument to the one we had for $y = ax^2$ in the previous chapter.

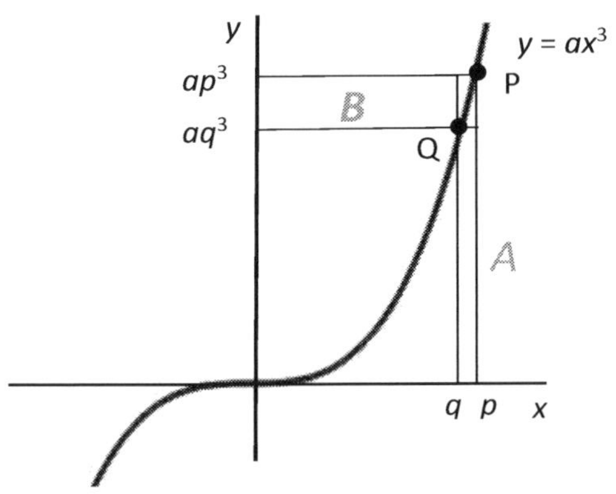

The area A under the curve between $x = p$ and $x = q$ is approximated by the rectangle with base $(p - q)$ and height aq^3.

The area B is approximated by the rectangle whose area is $q(ap^3 - aq^3)$.

The ratio of these areas is $\frac{B}{A} = \dfrac{q(ap^3 - aq^3)}{(p-q)aq^3}$.

We know we can cancel $(p - q)$ into $(ap^3 - aq^3)$ using our equation:

NOTES

$p^3 - q^3 = (p - q)(p^2 + pq + q^2)$

This allows us to simplify to $\frac{B}{A} = \frac{(p^2+pq+q^2)}{q^2}$

Now as Q merges with P, $q = p$ and we get $\frac{B}{A} = \frac{3p^2}{p^2} = 3$.

The 3:1 ratio means that the area under the curve is one quarter of the whole rectangular area.

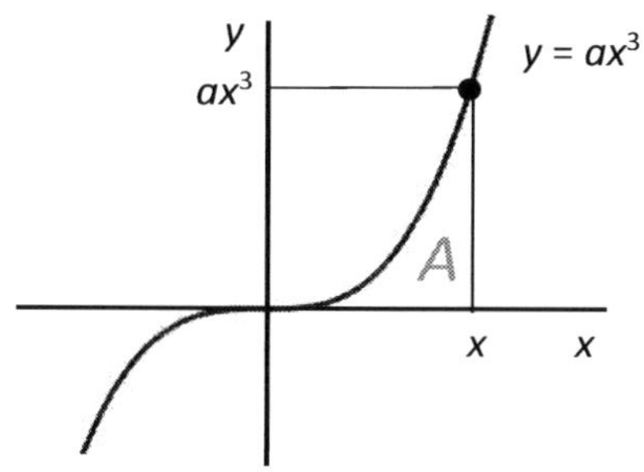

So we get $A = \frac{1}{4}ax^4$ and for a strip from $x = p$ to $x = q$ we take the difference of areas: $\frac{1}{4}ap^4 - \frac{1}{4}aq^4 = \frac{a}{4}(p^4 - q^4)$.

Chapter 10, Page 84

RATIO OF AREAS = *n*

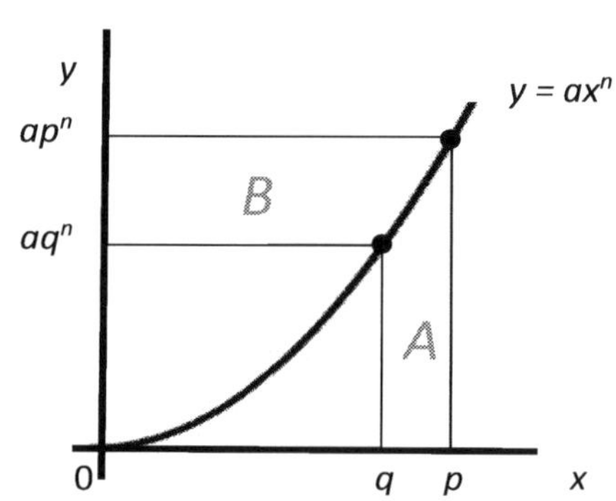

Note $(p - q)$ is a factor of $(p - q)^n$.

$$\frac{B}{A} = \frac{q(ap^n - aq^n)}{(p-q)aq^n} = \frac{(p^{n-1} + p^n q + \ldots q^{n-1})}{q^{n-1}}$$ in which the numerator has n terms.

Then the limit as q goes to p of $\left(\frac{(p^{n-1} + p^n q + \ldots q^{n-1})}{q^{n-1}}\right)$

= the limit as q goes to p of $\left(\frac{np^{n-1}}{p^{n-1}}\right) = n$.

Chapter 10, Page 84

RATIO OF GRADIENTS = RATIO OF AREAS = *n*

We know that if we take two close points on the curve $y = ax^n$ there are **two areas** associated with them: the area against the y-axis and the area against the x-axis.

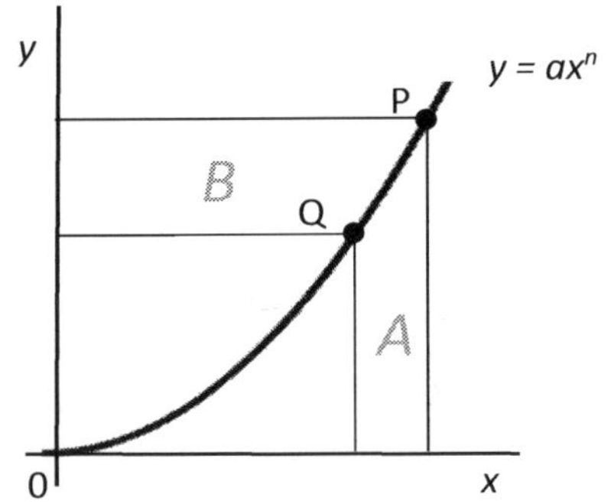

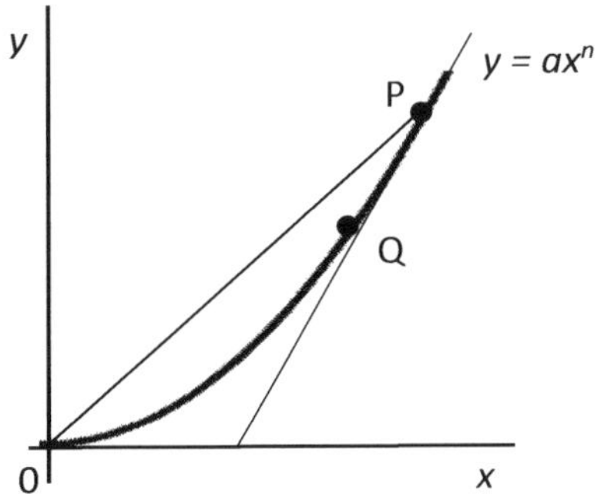

There are also **two gradients** associated with two such points: the gradient of the tangent to the curve at one of the points and the gradient of the line that joins one of them to the origin.

We are saying that the limiting ratio of areas equals the limiting ratio of gradients. And that this is also equal to n, where n is a rational number (fraction) greater than zero.

We have already seen that the limiting ratio of areas equals n.

What about the limiting ratio of gradients?

Let P have coordinates (p, y).
For simplicity we can let $p - q = u$,
and $ap^n - aq^n = v$,
as shown in the diagram.

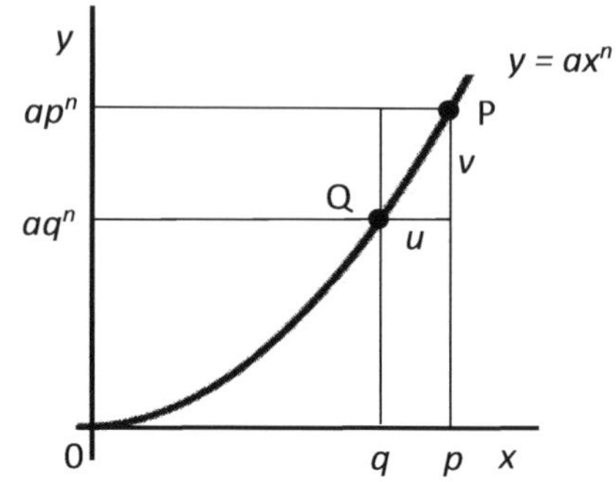

Then the **ratio of areas** is $\frac{vp}{uy}$.

And the limit of this ratio as $q \to p$

we know is n.

The gradient of the tangent at P is approximated to $\frac{v}{u}$.

And the gradient of OP is $\frac{y}{p}$.

The **ratio of these gradients** is $\frac{v/u}{y/p} = \frac{vp}{uy}$. Which is also the ratio of areas.

Now as Q approaches and merges into P, the ratio of areas becomes n.
And also the approximate gradient $\frac{v}{u}$ becomes an accurate gradient.
So the ratio of gradients must also become equal to n.

Therefore the limiting ratio of areas equals the limiting ratio of gradients and these are also equal to n.

<u>**CHAPTER 11, Page 90**</u>

The Integration Symbol

An area can be found by summing a large number of vertical strips, and this explains the symbols used in integration:

$$\int y \, dx$$

$\int$ is an elongated S standing for 'summation' and $y\,dx$ is the area of a rectangle of height y and width dx.

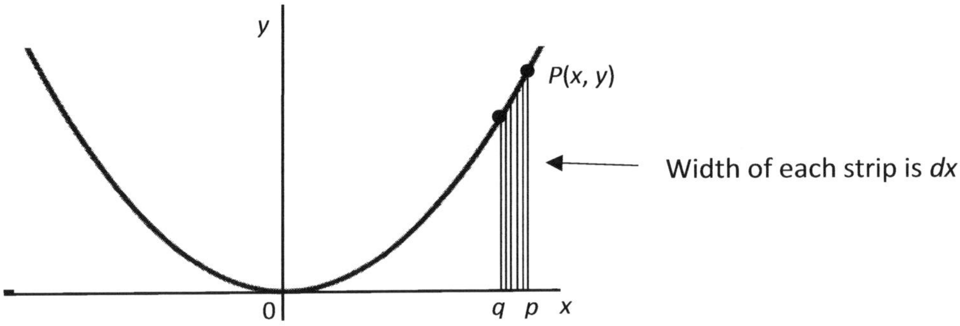

So $\int y\, dx$ means the sum of the areas of an infinite number of such strips.

ANSWERS

CHAPTER 1

Exercise A **a** 27 cm

b

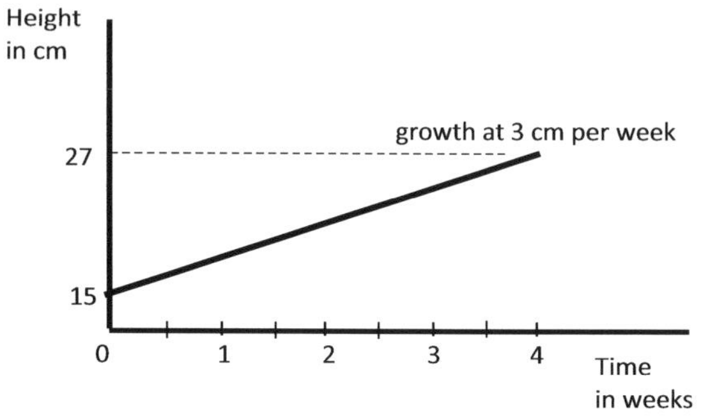

Exercise B The length of *BC* becomes zero as *B* merges into *C*.

Exercise C **d**

x	-4.5	-3.5	-2.5	-1.5	-0.5	0.5	1.5	2.5	3.5	4.5
y	20.25	12.25	**6.25**	**2.25**	0.25	**0.25**	**2.25**	6.25	**12.25**	20.25

f $\frac{25-1}{5--1} = 4$ **g** $\frac{25-4}{-5--2} = -7$

Exercise D **a**

x	-1	0	1	2	3	4
y	0	-3	**-4**	**-3**	0	5

c

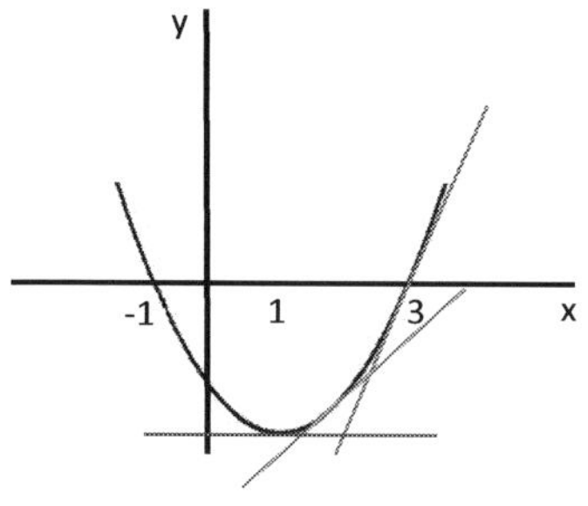

d 4, 2, 0

CHAPTER 2

Exercise A **a** 9 **b** 14 **c** 2 **d** -5

Exercise B **a** 13 **b** 20 **c** 6 **d** -11

CHAPTER 2 continued

Exercise C

a

Curve	x-coordinates	$\frac{Dy}{Dx}$ =
$y = 2x^2$	$x = 3$ and $x = 4$	**14**
$y = 3x^2$	$x = 3$ and $x = 4$	**21**
$y = 4x^2$	$x = 7$ and $x = 5$	**48**
$y = 5x^2$	$x = 9$ and $x = 0$	**45**
$y = -2x^2$	$x = 5$ and $x = 2$	**-14**
$y = 3x^2$	$x = 7$ and $x = -4$	**9**

b $\frac{Dy}{Dx} = a(p + q)$.

c If $y = ax^2$ and $x = p$ and $x = q$ are the two x-coordinates, then we have P(p, ap^2) and Q(q, aq^2).

So $\frac{Dy}{Dx} = \frac{ap^2 - aq^2}{p - q} = \frac{a(p^2 - q^2)}{p - q} = \frac{a(p - q)(p + q)}{p - q} = a(p + q)$.

Exercise D

a

Curve, $y = ax^2 + bx + c$	**x-coordinates p and q**	**$\frac{Dy}{Dx}$ =**	**$a(p + q)$**
$y = 2x^2 + 3x + 4$	$x = 5$ and $x = 1$	**15**	**12**
$y = 2x^2 + 3x + 5$	$x = 4$ and $x = 1$	**13**	**10**
$y = 3x^2 + x - 4$	$x = 3$ and $x = 1$	**13**	**12**
$y = 4x^2 + 3x + 2$	$x = 5$ and $x = 2$	**31**	**28**
$y = 5x^2 + 6x + 8$	$x = 3$ and $x = 0$	**21**	**15**

b $\frac{Dy}{Dx} = a(p + q) + b$.

CHAPTER 2, Exercise D continued

c

Curve, $y = ax^2 + bx + c$	x-coordinates p and q	$\frac{Dy}{Dx} = a(p + q) + b$
$y = 3x^2 + 4x + 1$	$x = 4$ and $x = 7$	**37**
$y = -2x^2 + 3x - 4$	$x = 5$ and $x = 3$	**-13**
$y = 4x^2 + 5$	$x = 4$ and $x = 1$	**20**
$y = x^2 - 2x + 15$	$x = 9$ and $x = -8$	**-1**

d Changing c will only move the curve up or down, so the gradient will not be changed whatever value c has.

e Start with the general parabola $y = ax^2 + bx + c$ and the two x-coordinates p and q.

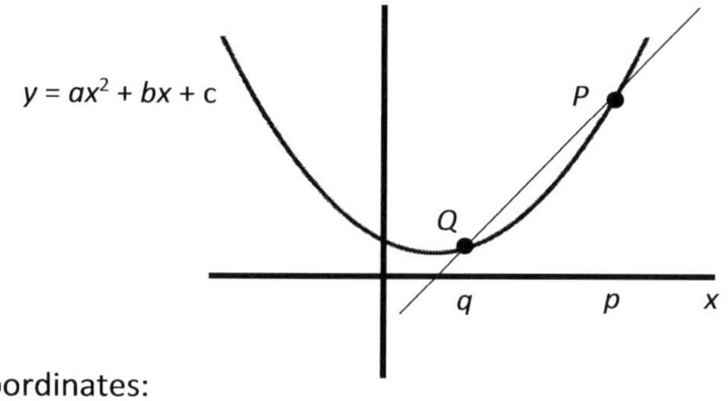

First find the y-coordinates:

If $x = p$ then $y = ap^2 + bp + c$, and
if $x = q$ then $y = aq^2 + bq + c$.

So the secant gradient is $\frac{Dy}{Dx} = \frac{(ap^2 + bp + c) - (aq^2 + bq + c)}{p - q}$

Here we see that c cancels out, and we can put the squared terms together to get:

$$\frac{(ap^2 - aq^2) + (bp - bq)}{p - q} = \frac{a(p^2 - q^2) + b(p - q)}{p - q}$$

and remembering that $(p^2 - q^2) = (p - q)(p + q)$, we can cancel $(p - q)$ to get:

$\frac{Dy}{Dx} = a(p + q) + b$.

CHAPTER 3

Exercise A

a 4 **b** 6 **c** 8

d Copy and complete the table below for the graph of $y = x^2$.

x	-5	-4	-3	-2	-1	0	1	2	3	4	5
y	**25**	**16**	**9**	**9**	**1**	0	1	4	9	**16**	**25**

g

x	-4.5	-3.5	-2.5	-1.5	-0.5	0.5	1.5	2.5	3.5	4.5
y	20.25	12.25	**6.25**	**2.25**	0.25	**0.25**	**2.25**	6.25	**12.25**	20.25

j You will find no such point

Exercise B

a 34 **b** -1 **c** 16 **d** 11

e 13 **f** -8

Exercise C

a 8 **b** -13 **c** 0 **d** 15 **e** 17 **f** 33

Exercise D

a $6x + 4$ **b** $-14x + 1$ **c** $-9 + 6x$

d $18x + 16$ **e** $2px + q$ **f** $2ax + 4$

g $2bx + 4$ **h** $2bx + c$ **i** -3

j 4, 2, 0

CHAPTER 3 continued

Exercise E

a (2, 2) **b** (0, 4) **c** $(-\frac{1}{2}, 17\frac{3}{4})$

d $(-\frac{1}{2}, -4\frac{1}{4})$ **e** (-6, 17) **f** (2, 20)

Exercise F

a $\frac{dy}{dx} = 2x + 41$ **b** $\frac{dy}{dx} = 4x - 1$ **c** $\frac{dy}{dx} = -x$

d $\frac{dy}{dt} = 6t + 1$ **e** $\frac{dN}{dy} = 6y$ **f** $\frac{dx}{dt} = -10t$

g $\frac{dy}{dx} = 2x + 1$ **h** $\frac{dT}{dx} = 2x + \frac{2}{3}$ **i** $\frac{dA}{dr} = 2\pi r$

Exercise G

a $8x - 1$ **b** $2x$ **c** $-5 - 8x$

d $-2x$ **e** $2x + \pi$ **f** $4x$

CHAPTER 4

Exercise A

a

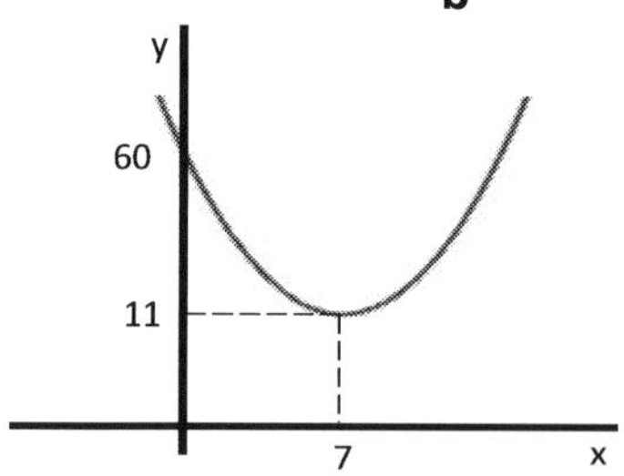

b

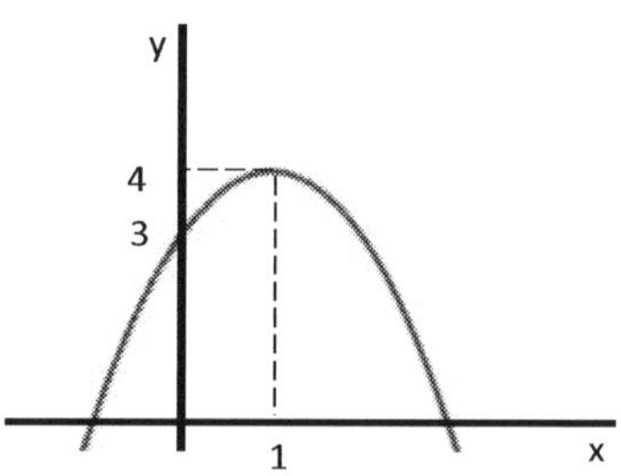

CHAPTER 4 continued

c

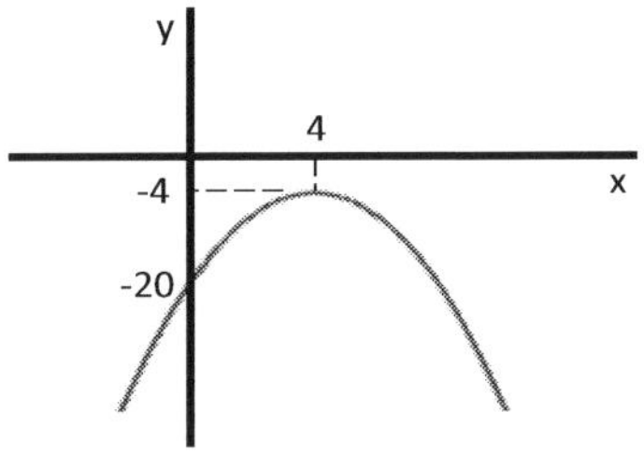

d

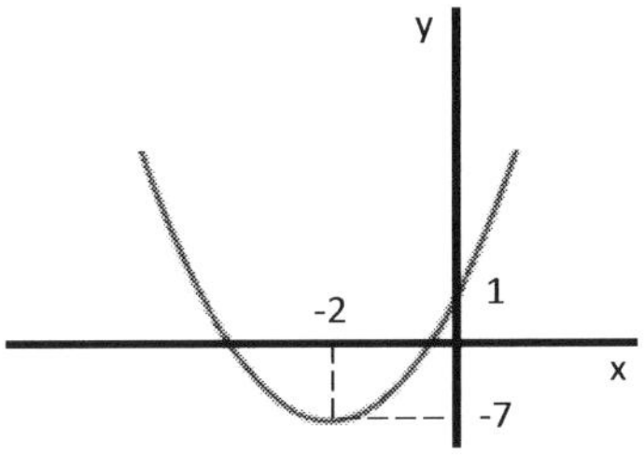

e

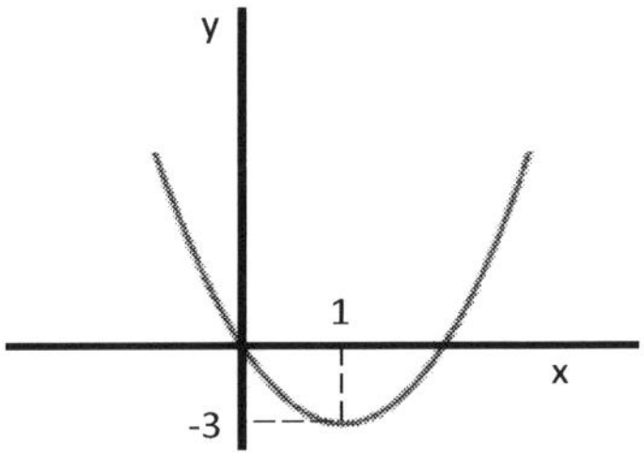

f

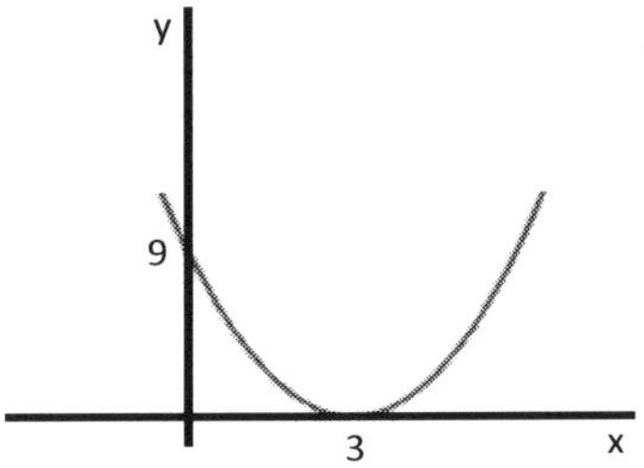

Exercise B

a 4 **b** -23 **c** -11 **d** 27

Exercise C

a 9 sec, 81 m **b** n = 50, cost = \$120 **c** x = \$10, P = \$50,000

Exercise D

4 feet by 8 feet

CHAPTER 5

Exercise A

b $(5^3 - 3^3) = 98$, $(5 - 3)(5^2 + 5\times3 + 3^2) = 98$

c $p^5 - q^5 = (p - q)(p^4 + p^3q + p^2q^2 + pq^3 + q^4)$

CHAPTER 5, Exercise A continued

d The power is **equal to** the number of terms

e The power of the first term is **1 more** than the power on the first term in the right-hand bracket

Exercise B

a, b

Equation	Secant Gradient	Tangent Gradient	Put $p = x$
$y = x$	1	1	1
$y = x^2$	$p + q$	$2p$	$2x$
$y = x^3$	$p^2 + pq + q^2$	$\mathbf{3p^2}$	$\mathbf{3x^2}$
$y = x^4$	$p^3 + p^2q + pq^2 + q^3$	$\mathbf{4p^3}$	$\mathbf{4x^3}$
$\mathbf{y = x^5}$	$\mathbf{p^4 + p^3q^2 + p^2q^3 + pq^4 + q^5}$	$\mathbf{5p^4}$	$\mathbf{5x^4}$

c 48 (since $3x^2 = 48$ when $x = 4$)
d 32

e The pattern is: the coefficient is the same as the power, and the power is reduced by 1

f $7x^6$ **g** $20x^{19}$ **h** nx^{n-1}

Exercise C

a $20x^3$ **b** $9x^2$ **c** $56x^6$ **d** $6x^5$ **e** $10x$ **f** 1

g $2x^3$ **h** $6x$ **i** 9 * **j** $60x^3$ **k** $36x^{11}$ **l** 0

* Note $x^0 = 1$ (this is explained in Chapter 7)

m If $y = ax^3$ then we have the 2 points as (p, ap^3) and (q, aq^3).

Then $\frac{dy}{dx} = \frac{ap^3 - aq^3}{p - q} = \frac{a(p^2 + pq + q^2)(p - q)}{p - q} = a(p^2 + pq + q^2)$.

And when $q = p$ then $\frac{dy}{dx} = a(3p^2) = 3ax^2$.

CHAPTER 5 continued

Exercise D

a $18x^5 + 20x^3$

b $-3x^2$

c $\frac{dy}{dx} = 12x^3 + 8x - 6$

d $\frac{dy}{dx} = 5 - 6x^2$

e $5x^4 - 3x^2 - 1$

f $x - \frac{3}{2}x^2$

g $39x^2 + 2x^3$

h $18x - 1 + 27x^2$

i $\frac{dy}{dx} = 0$

j $\frac{dy}{dx} = 50 - 100x$

k Secant gradient = $\frac{p^2+5p-q^2-5q}{p-q} = \frac{p^2-q^2+5(p-q)}{p-q} = p + q + 5 \rightarrow 2x + 5$

Exercise E

a $160x^3 + 60x^2$

b $2 - 6x$

c $6x - 1$

d $90x^4$

e -2

f 0

CHAPTER 6 Negative Powers

Exercise A

a

$x =$	-4	-2	-1	-½	-¼	¼	½	1	2	4
$y =$	-¼	**-½**	**-1**	**-2**	**-4**	4	**2**	**1**	**½**	**¼**

b -0.25

c -4

d -0.25

e -4

f Given points $P(p, \frac{1}{p})$ and $Q(q, \frac{1}{q})$, we get:

Secant gradient = $\frac{\frac{1}{p} - \frac{1}{q}}{p-q} = \frac{\frac{q-p}{pq}}{p-q} = -\frac{1}{pq}$.

And as $q \rightarrow p$ this becomes $-\frac{1}{p^2}$ which is $-\frac{1}{x^2}$.

CHAPTER 6, Exercise B continued

a -¼ **b** -4 **c** -¼ **d** -4

Exercise C

a

$x =$	-4	-2	-1	$-\frac{1}{2}$	0	$\frac{1}{2}$	1	2	4
$y =$	$\frac{1}{16}$	$\mathbf{\frac{1}{4}}$	1	**4**	∞	**4**	**1**	$\frac{1}{4}$	$\mathbf{\frac{1}{16}}$

e -2, -¼ **f** ¼ (by symmetry) **g** $y = x^{-2}$, so $\frac{dy}{dx} = -2x^{-3} = \frac{-2}{x^3}$

h Secant gradient = $\frac{p^2 - q^2}{\frac{1}{p} - \frac{1}{q}} = \frac{(p+q)(p-q)}{\frac{q-p}{pq}} = -pq(p+q)$

Putting $q = p$ we get: $-2p^3$.

And since $x = \frac{1}{p}$ we get $\frac{dy}{dx} = \frac{-2}{x^3}$.

Exercise D

a $\frac{-6}{x^3}$ **b** $\frac{-1}{3x^2}$ **c** $\frac{-21}{x^4}$ **d** 4 **e** -9

CHAPTER 7 Fractional Powers

Exercise A

a

$y =$	-3	-2	-1	0	1	2	3
$x =$	9	**4**	1	0	1	**4**	**9**

c

$y =$	-2.5	-1.5	-0.5	0.5	1.5	2.5
$x =$	6.25	**2.25**	0.25	0.25	**2.25**	6.25

CHAPTER 7, Exercise A continued

d 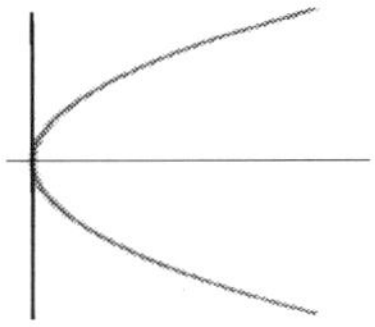

e ½, -½

f ¼, -¼

g Finding the Gradient formula for $y^2 = x$

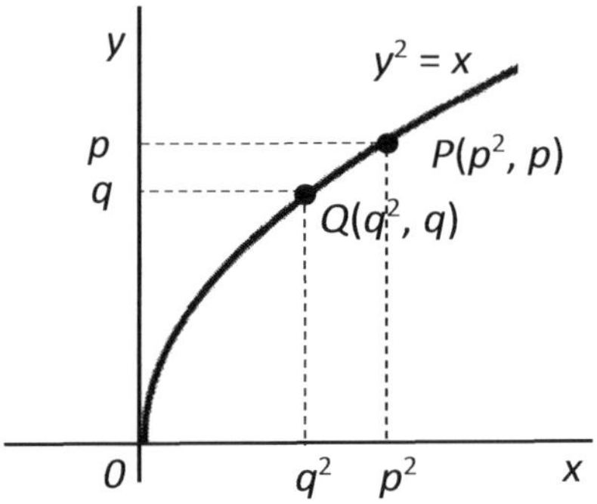

Using the secant-tangent method from before to find the gradient, we take two points on the curve, get the gradient of the line joining them and then allow one point to approach and merge into the other point.

We take a point, P, on the curve, and if its coordinates are (p^2, p) then we can see it will always lie on the curve because for (p^2, p) we find that $y^2 = x$.

Now we can see that the gradient of PQ is $\dfrac{p-q}{p^2-q^2}$.

Here we can cancel $(p - q)$ to get $\dfrac{1}{p+q}$.

Now as $Q \to P$ we get $\dfrac{1}{p+q} \to \dfrac{1}{2p} = \dfrac{1}{2y}$.

Or, since $p^2 = x$ we can replace p with $\pm\sqrt{x}$ to get:

$$\frac{dy}{dx} = \pm\frac{1}{2\sqrt{x}}.$$

CHAPTER 7, Exercise B

a ½, -½ **b** your answer should be the same or nearly the same

c ¼, -¼ **d** your answer should be the same or nearly the same

Exercise C

a

y =	-27	-8	-1	0	1	8	27
x =	9	4	1	0	1	4	9
y^2 =	**729**	**64**	**1**	**0**	**1**	**64**	**729**
x^3 =	**729**	**64**	**1**	**0**	**1**	**64**	**729**

e

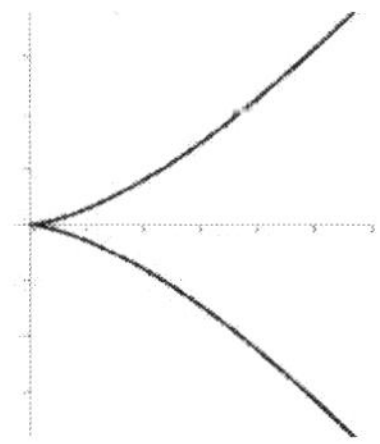

Exercise D

a ±3 **b** You should get about ±3

c ±1.5 **d** You should get about ±1.5

e You should get approximately ±2

f Starting with $P(p^2, p^3)$ and $Q(q^2, q^3)$ the

y-differences divided by the x-differences = $\dfrac{p^3 - q^3}{p^2 - q^2} = \dfrac{p^2 + pq + q^2}{p + q}$.

Putting $q = p$ we get $\dfrac{3p^2}{2p} = \frac{3}{2}\,p = \pm\frac{3}{2}x^{\frac{1}{2}}$ (since $p = \pm x^{½}$).

So $\dfrac{dy}{dx} = \pm\frac{3}{2}x^{\frac{1}{2}}$.

CHAPTER 7, Exercise E

a $\frac{5}{2}x^{\frac{3}{2}}$ **b** $5x^{\frac{3}{2}}$ **c** $4x^{-\frac{3}{5}}$ **d** $\frac{2}{3}x^{-\frac{1}{3}}$

e $\frac{1}{3}$ **f** $\frac{2}{5}$ **g** ± 20

CHAPTER 8 Ratio of Gradients

Exercise A

d gradients are 3 and 1.5 **e** you will find no such point

Exercise B

a 2 **b** 5 **c** -1 **d** 5/2

e 2/3 **f** 1/2 **g** -2

Exercise C

g see the diagram opposite for example

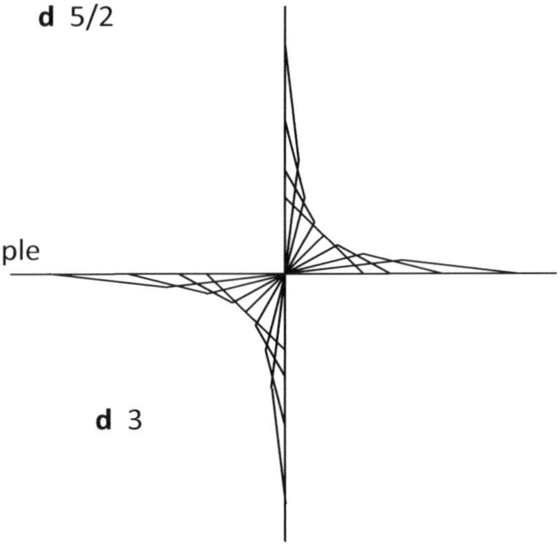

Exercise D

a -1 **b** -5 **c** -2/27 **d** 3

e 2/9 **f** -2/3 **g** -3/2

CHAPTER 9 Area Under a Parabola

Exercise A

a $\frac{1}{3}$ **b** 72 **c** $\frac{1}{24}$

d $\frac{8}{3}$ **e** $\frac{1000}{3}$ **f** 18 **g** $\frac{250}{3}$

i Intersections of $y = x$ with $y = x^2$ are at (0, 0) and (1, 1). $A = \frac{1}{2} - \frac{1}{3} = \frac{1}{6}$.

CHAPTER 9 continued

Exercise B

a $\frac{98}{3}$ **b** $\frac{152}{3}$ **c** $\frac{7}{3}$ **d** $\frac{56}{3}$ **e** $\frac{38}{3}$ **f** $\frac{74}{3}$

Exercise C

a $\frac{19}{3}$ **b** 21 **c** $\frac{19}{3}$ **d** $\frac{52}{3}$

Exercise D

a 7 **b** $\frac{38}{3}$ **c** 105 **d** $\frac{70}{3}$ **e** $\frac{38}{3}$ **f** $\frac{7}{3}$

g $\frac{7}{3}$ **h** 7

CHAPTER 10

Exercise A

b 30 **c** 48 **d** 40 **e** $\frac{5}{4}$

Exercise B

a 8 **b** 5 **c** 24 **d** $\frac{35}{2}$ **e** $\frac{248}{3}$ **f** $\frac{185}{3}$

Exercise C

a $\frac{32}{3}$ **b** $\frac{82}{3}$ **c** 6 **d** $\frac{9}{2}$

Exercise D

a $\frac{255}{4}$ **b** $\frac{81}{2}$ **c** $\frac{31}{5}$ **d** $A = \frac{a}{n+1}(p^{n+1} - q^{n+1})$ **e** $\frac{96}{5}$

Exercise E

a $\frac{112}{3}$ **b** 20 **c** $\frac{291}{4}$

CHAPTER 11

Exercise A

a C **b** A

Exercise B

a $2x^3 + c$ **b** $2x^4 + c$ **c** $3x^5 + c$

Exercise C

a $\frac{5}{3}x^3 + c$ **b** $5x^4 + c$ **c** $x^5 + c$ **d** $\frac{1}{4}x^4 + c$

e $\frac{1}{5}x^5 + c$ **f** $\frac{2}{7}x^7 + c$ **g** $\frac{1}{8}x^4 + c$ **h** $\frac{1}{7}x^2 + c$

Exercise D

a $4x^3 + c$ **b** $3x^4 + c$ **c** $\frac{12}{5}x^5 + c$ **d** $6x^2 + c$ **e** $12x + c$

f $\frac{1}{3}x^3 + c$ **g** $5x^3 + c$ **h** $\frac{1}{8}x^4 + c$ **i** $\frac{3}{8}x^8 + c$ **j** $2x + c$

k $ax + c$ **l** $\frac{a}{2}x^2 + c$ **m** $\frac{a}{3}x^3 + c$

n $y = x^3 + c$ **o** $y = x^9 + c$ **p** $y = 2x^7 + c$

q $y = \frac{2}{3}x^6 + c$ **r** $y = \frac{17}{2}x^2 + c$

Exercise E

a $2x^3 + 4x^2 + 10x + c$ **b** $3x - 3x^2 + c$ **c** $\frac{1}{3}x^3 + \frac{1}{2}x^2 + x + c$

d $\frac{5}{2}x^2 - 4x + c$ **e** $y = x^5 - 6x^3 + c$

Exercise F

a $y = x^2 + 3x + 3$ **b** $y = x^3 + 2x^2 - 5x + 3$ **c** $y = 2x - x^3 - 1$

CHAPTER 12

Exercise A

a

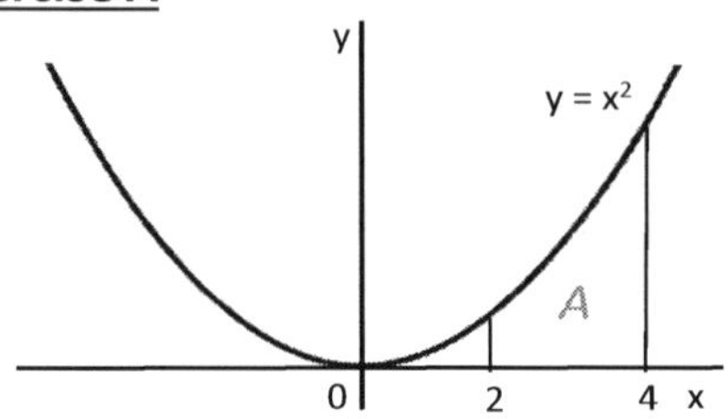

b $\frac{56}{3}$ **c** $\int_3^4 x^2 dx$ **d** 117

e $\int_q^p ax^n dx = \left[\frac{a}{n+1}x^{n+1}\right]_q^p = \frac{a}{n+1}p^{n+1} - \frac{a}{n+1}q^{n+1} = \frac{a}{n+1}(p^{n+1} - q^{n+1})$.

Exercise B

a 12 **b** $\frac{74}{3}$ **c** $\frac{74}{3} + 2 = \frac{80}{3}$

CHAPTER 13

Exercise A

$t =$	0	1	2	3
$s =$	0	4	2	-12

The particle is moving to the right, then leftwards.

Exercise B

a 27 m/s, $t = 0.3$ s
b $v = -12$ m/s
c $v = -12$ m/s, $t = 2$ s

Exercise C

a $v = 3t^2 - 8t - 5$, $a = 6t - 8$
b 5 m, 20 m/s, 42 m/s^2
c $a = 24t + 7$, $v = 10$ m/s, $a = 7$ m/s^2
d $v = 20$ m/s, $t = 2$ s, $h = 20$ m (at the greatest height the velocity is zero)

CHAPTER 13 continued

Exercise D

a $s = 2t^3 + 4t^2 - t + c$, $a = 12t + 8$
b $v = 6t^2 + 7$, $s = 2t^3 + 7t$
c $t = 4$ s, $s = -20$ m

CHAPTER 14

Exercise A

a $2y + 3y' = 8x + 22$ **b** $y + 6y' = 2x + 9$

c $3y - y' = 21x - 4$ **d** $4y + y' = 20x + 5$

e $5y + 3y' = 5(4x - 3) + 3\times4 = 20x - 15 + 12 = 20x - 3$

Exercise B

a $y = 5x + 1$ **b** $y = x + 3$

c $y = 3x + 2$ **d** $y = 3x + 1$

e $y = x + 3$ **f** $y = x + 1$

g $y = 5x - 1$ **h** $y = 3x + \frac{1}{2}$

INDEX

Printed in Dunstable, United Kingdom

70863011R00088